AF553902

ENVIRONMENT CONCERNS AND STRATEGIES

ENVIRONMENT CONCERNS AND STRATEGIES

A.K. Shrivastava
Chairman-cum-Chief Executive "CTREE"

In association with COUNCIL FOR TRAINING & RESEARCH IN ECOLOGY & ENVIRONMENT

A P H PUBLISHING CORPORATION
7 ANSARI ROAD, DARYA GANJ
NEW DELHI-110 002

Supported by: **All India Council for Technical Education (Ministry of Human Resource Development, Government of India), New Delhi**

Published by
S.B. Nangia
A P H Publishing Corporation
7, Ansari Road, Daryaganj
New Delhi 110002
Ph.: 23274050
E-mail : aphbooks@gmail.com

2026

₹2995/-

ISBN 81-313-0162-1

Printed at
Balaji Offset
Navin Shahdara, Delhi 110032

PREFACE

Environment is, of course, crucial to the poorest people of our country, providing them with food, water, energy, employment and raw materials. But the poverty in which they live or some times they forced to live, reciprocally forces them to use these essential resources unsustainably. On the other hand, a long term over consumption and commercialization of resources has had a huge advice impact on the environment that pollution, waste dumps and global warming.

In fact the richer countries as well the poor one, can be made responsible for environmental loss and degradation. But the responsibilities of richer countries should have much more concern with the alarming situation. The developed countries like India need to show genuine commitment to changing unsustainable patterns of natural resource's consumptions and far greater efficiency in resource use. No doubt, the summits on development and sustainability, earth and environment has played a significant role in pressurizing the governments and development proponents to ensure that development really can be made lasting and sustainable for the world's poorest people, and for their children and further generations.

Water, sanitation, energy, health, agriculture, trade, non-voilence, population and biodiversity have already been identifies by the global environment agencies as key areas where concrete results can be achieved. The government of India has also designed such area and pattern of services but it not reflects the vision or mission of the development and sustainable development. It requires well spelt and coordinated approach to resolve the environment and sustainable development issues.

ACKNOWLEDGEMENT

The author's thank are due to the All India Council for Technical Education, Ministry of Human Resource Development, Government of India, the encouraging attitude of whom, helps the author and the CTREE to get this book published. The author would like to thank CTREE's participants to work out on the economic aspects of sustainable development in our country and abroad.

And, finally the author would like to express his thanks to the decision makers and planners of our country, who has started planning on sustainable development practices.

A.K. Shrivastava

CONTENTS

Preface v
Acknowledgement vii
1. Introduction 1
2. Economic Development Activities under UN Programmes 5
3. Local Economic Development 9
4. Lessons Learned 23
5. Economics Aspects of Sustainable Development in India 27
6. Social Aspects of Sustainable Development in India 159
7. Science: Environment: Global change: Acivists and NGOs 229
8. Agriculture and Biotechnology 255
9. Energy and Climate Change 257
10. Globalization 261
11. Depletion of Drinking Water: A Case Study of Coca-cola 265
Index 271

1

INTRODUCTION

Most countries in Central America have, at one point or another, suffered from civil conflict. Social inequity and a lack of democracy are usually seen as the causes of these conflicts. The issue of land ownership is particularly important in this respect. The heavily skewed land distribution had the effect of excluding large parts of the rural population from sharing in the economic benefits of agriculture. Guatemala is a good example. According to the World Bank, the agricultural census of 1979 indicated that 2.5 percent of Guatemala's 5.3 million farms controlled 65 percent of agricultural land, while only 16 percent of the land was cultivated by 88 percent of the smallest farms (World Bank, 1995). The World Bank considers the persistence of this inequality an important cause of the high incidence of poverty in Guatemala. Account also needs to be taken of the fact that industrial development in the region is limited to a few urban areas and thus has benefitted only a small percentage of the population.

In addition, socio-economic differences have traditionally been exacerbated by violence on the part of mostly military governments, making peaceful change almost impossible. This situation eventually led to armed conflict when opposition groups began to organize themselves as guerrilla movements in Nicaragua, El Salvador and Guatemala.

In 1979 the *Frente Sandinista de Liberación Nacional* in Nicaragua, which enjoyed wide popular support, overthrew the dictatorship of the Somoza family, seeking social and economic reform. Its ability to implement reforms however, was seriously undermined by violent strife provoked by the so-called *contras*, which received support from the United States. This lasted until 1990 when Mrs. Violeta Chamorro of the opposition party UNO gained the presidency after general elections. Small guerrilla groups (*recontras* and *recompas*) remained active until final peace agreements were signed in 1994.

The 1970s saw the emergence of a guerrilla movement in El Salvador, after it became clear that peaceful protest against the lack of democracy did not have any results. A number of different guerrilla groups expanded rapidly and in reaction army and paramilitary death squads became more active in their attacks on opposition groups. At the end of 1980 the different guerrilla groups unified into the *Frente Farabundo Martí para la Liberación Nacional* (FMLN) and started a guerrilla war which was to last twelve years. Although the FMLN never succeeded in instigating a nation-wide insurrection, considerable parts of the country, particularly the North and East, suffered serious violent conflict, with the departments of Chalatenango and Morazán being practically controlled by the FMLN.

In Guatemala violent confrontations between the armed forces and indigenous groups started even earlier than in Nicaragua and El Salvador. Since 1954 Guatemala was governed by a succession of right-wing civilian and military governments which suppressed any left-wing and centrist political parties and trade unions. This situation led to the formation of guerrilla groups in the 1960s. Confrontations between guerillas and the army became ever more frequent and violent during the 1980's. Large groups of civilians, especially among the Maya population suffered from harsh repression by the military. The departments of Huehuetenango and El Quiché were particularly affected. These are also the poorest regions of Guatemala. Peace negotiations started in 1990, but a peace accord was not signed until December 1996.

Although in Honduras guerrilla activities never turned into a civil war situation, large numbers of people suffered from violent repression on the part of military or military-controlled governments. Honduras was also confronted with large numbers of refugees from El Salvador and provided a staging area for Nicaraguan *contras* .

The peace process in Central America took off in 1987 with a peace treaty proposed by the president of Costa Rica, Oscar Arias. The treaty was signed by the leaders of the five Central American republics in Esquipulas, Guatemala. By signing this treaty governments committed themselves to start a dialogue with the guerrilla groups and initiate a process of democratization.

A subsequent summit in Tela, Honduras led to the agreement that the anti-Sandinista *contras* should demobilize and that El Salvador's FMLN guerrillas should cease fire and negotiate with the government.

Although the US administration was unwilling to stop supplying aid to the *contras* until elections had been held in Nicaragua, the defeat of the Sandinistas in the February 1990 elections finally heralded the disbanding of the rebel forces in Nicaragua. The Salvadorian government and the FMLN finally agreed to end El Salvador's civil war through a peace accord signed in Mexico in January 1992. Since then the FMLN has been completely demobilized and recognized as a legitimate political party, participating in all elections since 1994. A Truth Commission has reported on human rights abuses during the conflict while the Salvadorian armed forces have been reduced in size, purged of a number of officers accused of human rights violations and relieved of responsibility for internal security. The old *Policía Nacional* was replaced by a civilian police force in December 1994.

In Guatemala, during the government of the former human rights ombudsman Ramiro de León Carpio, a timetable for discussions was agreed upon in March 1994, with the intention of achieving a firm and lasting peace settlement in December of the same year. This proved to be too optimistic, but in 1994 the Guatemalan government agreed to the demands of the *Unidad Revolucionaria Nacional Guatemalteca* (URNG) for human rights verification and the establishment of a Truth Commission. A UN mission (MINUGUA) arrived in November 1994. The two sides were unable to find sufficient common ground to reach a final peace accord in December, but they signed an agreement on indigenous peoples' rights and identity in March 1995 and an accord on socio-economic issues and the land question in May 1996. The final peace accord was signed in December 1996.

The peace process in Central America received wide backing from the international community. Initially support came from the Contadora group, which supported the initiatives of President Arias, while in May 1988 the United Nations General Assembly approved the Special Economic Cooperation Plan for Central America.

2

ECONOMIC DEVELOPMENT ACTIVITIES UNDER UN PROGRAMME

BACKGROUND

In the context of the Special Economic Cooperation Plan for Central America, the United Nations launched two major initiatives. One was the International Conference on Central American Refugees (CIREFCA), for which UNHCR took responsibility. The other was the Development Programme for Displaced Persons, Refugees and Returnees in Central America (PRODERE). The Government of Italy decided to funding PRODERE through UNDP to the tune of $115 million to facilitate the reintegration of those who had been forced to seek refuge in neighbouring countries or more stable areas of their own country. In view of its innovative and multi-disciplinary approach UNDP charged UNOPS with the execution of PRODERE. UNOPS in turn subcontracted ILO, UNHCR and WHO for specific components of PRODERE. The programme, approved in 1989, started operations in 1990. It was mainly active in El Salvador, Guatemala, and Nicaragua but also undertook some activities in Belize, Costa Rica and Honduras, countries which had offered asylum to those fleeing conflict.

PRODERE concentrated its activities in regions averaging 250,000 inhabitants. These were either areas hard hit by violent conflict or those hosting great numbers of refugees. For the most part, the areas covered followed departmental borders and included municipalities, villages and hamlets. The areas selected were characterized by high rates of social and economic exclusion. Central government institutions were mostly unrepresented, most basic infrastructure facilities had been destroyed and the areas were under military control. People who fled their homes because of ongoing conflicts were often held in transit camps, such as Xemamatze in Ixil, Guatemala, and then forced to resettle in communities controlled by the military. At the same time, central governments were undergoing a process of restructuring and reduction, thus weakening even more the government's capacity to be a partner in the implementation of programmes for returnees.

Most people in the areas covered by PRODERE used to rely mainly on self-employment in small, family farms and to a lesser extent in non-agricultural micro-enterprises. Salaried employment in the private sector was practically non-existent. The already high pre-conflict levels of poverty had become even more dramatic because of the fact that many farmers had abandoned their land and crops during the conflict. The land tenure issue, one of the root causes of the conflict in the first place, was further complicated because of numerous disputes over small holdings. Social coherence was also affected since almost every family counted one or more deceased, disappeared or refugees amongst its members. Widows, orphans and war victims of all ages were highly visible in many villages.

2.2 STRATEGY

PRODERE set itself the following objectives:

- promoting human rights;
- building a consensus around development issues;
- facilitating the reintegration of returnees;
- restoring basic services in such areas as health, education and housing;
- reactivating the local economy.

In order to achieve these objectives, PRODERE developed a strategy rather different from traditional technical cooperation programmes. Instead of intervening at the central government level or targeting one or more specific groups, PRODERE adopted a local development strategy based on a decentralized, integrated and bottom-up approach to development in limited geographical areas. The decision to adopt this strategy was based on the consideration that the civil strife in Central America affected particularly those regions that were poorest as a result of their neglect by national governments in the pre-conflict period. In fact this neglect was one of the main reasons for the outbreak of the conflict in the first place.

In addition it was felt that the transition towards a more democratic and participatory society should start at the community or municipality level since it is at this level that government policies and interventions have an immediate impact on the daily life of citizens and that citizen participation in decision making can be more easily achieved. The community level also offered PRODERE a chance to build a consensus of the local population around common interests following a period of conflict which had often pitted neighbour against neighbour. Furthermore the governments of Guatemala and Honduras had just decided to implement decentralization policies, which PRODERE could build upon.

An important consequence of the adoption of a local development strategy by PRODERE was its non-discriminatory approach, not favouring one group over another. For example, PRODERE activities for returnees would also include the population that had stayed behind during the conflict, while in the municipal technical committees promoted by PRODERE in Nicaragua ex-combatants from both sides joined mayors and producer associations to determine who would benefit from PRODERE investments and define a development plan.

A guiding principle behind PRODERE was the practical application of the Universal Declaration of Human Rights. The promotion and protection of human rights, apart from being a specific component of PRODERE, permeated many of its activities. PRODERE thus contributed to providing individuals with proper documentation and assisted in securing property titles. It promoted the organization of local branches of human-rights offices and facilitated the local administration of justice.

In the area of health care PRODERE promoted local health systems. Instead of dealing separately with each dimension of basic health care such as health care planning, infrastructure, mother and child care etc., an integrated plan was designed to improve health care in each area dealing with all technical, administrative, management and infrastructure issues involved. The health system centered around a central point, normally a hospital and usually covered several municipalities.

PRODERE also promoted departmental planning councils. These dealt with all local development issues in the area covered and aimed at a more integrated planning of local development. These councils considerably facilitated establishing the necessary linkages between the different subject areas covered by PRODERE such as health, education, local economic development, human rights and regional planning. This integrated approach was based on the conviction that social and economic development are necessarily interdependent. The planning councils thus managed the complex relation between different sectoral investments introducing consultation processes involving all concerned. They coordinated the plans formulated at municipal, departmental, regional and national levels, defined minimum development objectives and decided on the projects to be financed from national and international resources.

Initially, PRODERE had to overcome considerable mistrust on both sides. Each party in the conflict considered that the programme was supporting the other side. In El Salvador, Guatemala and Nicaragua the opposition considered the programme a government instrument to counter insurrection, while the military considered it as supporting the opposition. As a result PRODERE activities in each area were defined following a careful needs assessment, trying to enure that they covered the needs of the population as a whole. In many cases this meant that considerable time was spent to gain credibility as a neutral operation and to achieve at least a minimum consensus on what were considered the most urgent needs.

The most effective way for PRODERE to gain the confidence of both sides was to show concrete results. This often meant concentrating in the beginning on restoring basic services and infrastructure. In Nicaragua for instance, the programme organized various meetings with the local population with whom an emergency plan was designed. The plan focused on basic food production and with financial and technical assistance returnees could start to cultivate maize and beans again as well as recuperate abandoned coffee plantations in several municipalities. Access roads to the production zones were urgently repaired, provisional health care was provided and schools were rebuilt. It was only later that PRODERE started to organize the so-called municipal technical committees as instruments of agreement and decision making.

3

LOCAL ECONOMIC DEVELOPMENT

THE POST-CONFLICT ECONOMY

As described earlier, the post-conflict economy in rural areas of Central America was mainly a subsistence economy. Most people in the affected areas relied heavily on self-employment in small, family farms and to a lesser extent in informal non-agricultural micro-enterprises. Micro- and small enterprises faced the usual obstacles: lack of finance for investment, difficult market access, lack of information, lack of technical and basic management skills, resulting in low productivity and competitiveness and inhibiting their effectively linking up with the modern sector. They had also been negatively affected by the conflict itself. Returnees and ex-combatants faced tremendous difficulties to restart their productive activities as a result of the destruction of resources and infrastructure, the loss of perennial crops and the disappearance of village markets. In addition, the land property question had become more complicated because of numerous disputes over small-holdings between returning refugees and new settlers. This in turn resulted in a subsistence agriculture less diverse than before the conflict, leaving the farmers even less room for manoeuver than before. Displaced persons who settled in urban areas flooded the urban informal sector thus reducing even further the already meager income of existing informal sector operators.

THE LED METHODOLOGY

In this post-conflict context PRODERE introduced a local economic development (LED) strategy. It invited the ILO to implement this component of the programme in view of its experience and mandate in this area. LED aims at employment creation and income generation through an optimal use of human and other resources in a determined geographical area, the so-called "economic catchment area". An important aspect of LED is that groups traditionally excluded from the economic development process have a chance to become active members and beneficiaries of the local economy. A key principle of this approach is that the planning of resource utilization is done by the population itself through their own institutions. This means an emphasis on strengthening existing capacities through institution building, consensus and coordination as well as on providing new services. Usually this is achieved through the creation of a Local Economic Development Agency (LEDA). The LED methodology thus involves: consensus building; strengthening local capacities; a bottom-up, participatory approach; a strive for synergy; acting as catalyst; linking the local economy to the national and global economies; and raising public awareness.

Consensus can only be achieved through an effective participation of the local socio-economic actors concerned. This means that a process needs to be undertaken that brings together and links local actors across political lines and ensures a constructive exchange of ideas and opinions with the aim of designing policies for sustainable economic development in the area. This process is in itself an important part of the reconciliation and peace building process. Capacity building is essential in order to guarantee the technical sustainability of the initiatives to be developed. A bottom-up and participatory approach is required to mobilize to the fullest possible extent the local human potential, thus ensuring local "ownership" of the LED process. In order to achieve this it is important that concrete and visible results are achieved in a short period. Such results contribute to raising the level of motivation and awareness of the target population. An important way of ensuring quick results is to strive for synergy with other development efforts in the area. To ensure their longer term viability local initiatives also need to be linked to the national and global levels. It is therefore necessary to establish a dialogue with the relevant institutions at the national and international levels and inform them about local level initiatives. **Public awareness raising is important to show the importance of entrepreneurial initiatives for the local economy.**

LOCAL ECONOMIC DEVELOPMENT AGENCIES

Objectives

Local economic development agencies (LEDAs) aim at achieving a consensus among their members on the local economic development strategy most appropriate to the area. This consensus is usually arrived at following an analysis of the economic opportunities, constraints and potential of the area. LEDAs also provide technical and financial assistance to their members and others to help start, reactivate and strengthen enterprises, particularly those that have a potential for employment creation and are environmentally friendly. Since LEDAs facilitate a decision-making process based on the problems identified by the area's population itself, they can become a key instrument for the economic development of the area. LEDAs can also play an important role in channelling and coordinating international technical cooperation, which often intervenes in an area in a rather uncoordinated and arbitrary fashion.

Although in general LEDAs focus their activities mainly on the economic development of an area, in the case of PRODERE LEDAs were also key instruments in promoting consensus building, conciliation, democracy and popular participation. Many of the LEDAs set up in the framework of PRODERE, operate in geographical backward areas characterized by an 'institutional vacuum'. They were often one of the few, or even the only active institution in the area. Also, as compared with the Church or the local administration, LEDAs more easily gained acceptance as a neutral entity. Moreover, since some of the actors who were involved in the LEDAs also played an active role in health issues, education or general community development initiatives, some LEDAs found themselves involved in and functioned as catalysts for many other activities and events with a more social character. The main focus of the LEDAs however remained on economic development.

Geographical coverage

LEDAs vary in their geographical coverage from one or more municipalities to the level of a department. They also vary in terms of their composition. Although local in the context of PRODERE often meant provincial or departmental, the geopolitical unit was not the only criterion for the territorial definition of a LEDA. Others were cultural homogeneity, socio-economic coherence, income generation capacity (to improve the chances of financial sustainability) and possibilities for effective community participation.

A good example of the application of the above criteria is Guatemala where the regions Ixil and Ixcan are part the department of El Quiche. Ixcan however, is separated from the rest of the department by a mountain range reaching 4000 meters. Between Ixil in the highland and Ixcan in the lowland there is no cultural or socio-economic coherence, thus making effective community participation impossible. The municipality of Ixcan (being only one municipality) therefore created its own LEDA, while the Ixil LEDA extended its area of influence from the original Ixil-Triangle to other highland communities in eastern and southern Quiche.

It is also important to note that the word "local" in the title does by no means indicate that LEDAs restrict their activities to the local level. Very often they deal with institutions at the national level to solve specific problems, obtain funding or know-how.

With direct assistance from PRODERE LEDAs were established in Ixil and Ixcan in the department of El Quiché in Guatemala, in Morazán and Chalatenango in El Salvador, in Jinotega and Nueva Segovia in Nicaragua and in Intibuca and Ocotepeque in Honduras. In addition, through parallel bilateral interventions also funded by the Italian government, LEDAs were created in Granada and Leon in Nicaragua and three others in San Salvador. With funding from the EU another LEDA was established in Huehuetenango in Guatemala, while in Costa Rica an existing entity in Brunca which already was undertaking some activities similar to those of a LEDA was converted into a fully-fledged LEDA using its own resources and some support from PRODERE. Altogether these LEDAs serve an area of close to 35,000 square kilometers with over 3 million people (see map). Among them they have some 350 member institutions.

Organisation

The LEDAs supported by PRODERE were designed as membership organisations, made up of representatives of the public sector (including local administration and decentralized national government agencies) as well as of civil society (including peasant associations, cooperatives, private sector employers, workers' organizations and other non-governmental organizations). In this context, the role of PRODERE was limited to promoting the constitution of LEDAs and providing technical assistance. For the LEDAs to function properly, they had to be independent bodies, with a proper legal structure. It was therefore important that the local actors assume responsibility for the process of creating LEDAs right from the start. In order to ensure that the local community would not consider a LEDA as an institution imposed from the outside, but rather assume "ownership", the LEDAs were constituted only after an intensive, participatory process of promotion and preparation, involving ad-hoc committees on which most of the institutions mentioned above were represented. Ideally, LEDA membership should reflect the whole range of organizations and agencies active in the area, including local authorities, decentralized central government agencies and organized civil society. In practice however different LEDAs establish different membership criteria. One LEDA for instance did not accept farmers as members, since they were not organized. Many farmers however are part of other organizations which are LEDA member. Their interests are therefore represented indirectly

The constitution of a LEDA in Honduras

In October 1991, the municipality of San Marcos de Ocotepeque proposed to the *patronatos comunales* (traditional grass roots organizations at village level) of the 17 villages in the municipality to create a patronato superior", under presidency of the mayor and made up of the presidents of each *patronato comunal* which would be representative of the whole municipality.

Parallel to this initiative an inter-institutional coordination committee was created to coordinate the activities of the community, local government and other locally active institutions. In 1992, PRODERE stimulated the dialogue between municipalities and institutions by providing a coordination forum. In the same year a workshop was organized concerning on an integrated local development approach, in which public and private institutions, farmers organizations, cooperatives and the municipality of San Marcos participated. The meeting resulted in the creation of a municipal development organization to orient the development process and study the feasibility of the installation of a local economic development agency. After discussions with the productive sector at the grassroots level in the 7 municipalities of the Sensenti Valley, and with the mayors of the respective municipalities a local development agency was then created.

Overall conditions in Honduras were also favourable since new legislation on the functioning of municipalities foresaw greater decentralization and democratization. Following the example of San Marcos, PRODERE also promoted municipal development councils in the 6 other municipalities of the Sensenti valley. These councils included representatives of the local government as well as of the *patronatos comunales*. The process of promoting the councils and the LEDA was simultaneous and interlinked with the councils representing the grassroots level in the LEDA. At the end of 1992, the 7 councils agreed to the establishment of the LEDA, which was installed in January 1993. In addition to the councils a number of other public and private institutions, including farmers' organizations joined the LEDA.

In PRODERE the average cost of creating a LEDA was about $120,000. This amount includes not only the cost of equipping the LEDA but also the running costs for the first year of operations and the cost of training LEDA staff.

It is further worthwhile to note that in 1994 the 15 LEDAs mentioned above decided to create the Central American LEDA Network. This network has enabled the LEDAs to be represented at international meetings, exchange experiences, exploit trading opportunities, develop joint services, realize economies of scale, undertake joint training activities and mobilize external resources. National networks have also been set up in Guatemala, Honduras and El Salvador for similar purposes.

Activities

The first task of a LEDA after its constitution is to assess the area's potential in terms of the available natural, economic, human and financial resources and institutional presence. The outcome of this assessment then serves as basis for strategy planning and coordination of activities. All LEDA members should be involved in the assessment for which normally the SWOT (Strengths, Weaknesses, Opportunities, Threats) method is used. This method is often also used to anlyze the strengths and weaknesses of LEDA member organizations.

Following this assessment an investment opportunity study for the promotion of micro-and small enterprises is usually carried out. PRODERE developed a special method for this, the so-called *diagnostico preliminar de oportunidades de inversion* (preliminary investment opportunity analysis). This method is participatory and involves all sectors of the community concerned in the identification of economic needs and in the possibilities to satisfy them through entrepreneurial activities.

Both the SWOT analysis and the investment opportunity study are used to design an economic development plan. This is essential since the strategic planning concept is essential to the LED methodology. Local development should not be incidental or accidental. Spontaneous and natural evolution processes are important but it is even more important to ensure that local and external players focus their attention and energy on an analysis of the situation and then jointly develop a strategy and undertake the activities necessary to implement the strategy. LEDAs do not only function on the basis of their own strategic plan however. Their activities are also guided by an overall economic development for the area. This plan defines the long term strategic direction of the development of the local community, practical projects to be implemented, financial, material and technical support required to achieve the goals set and the involvement of individual LEDA members and others in its implementation.

As mentioned before, LEDAs aim at achieving synergy and complementing other ongoing activities in order to make an effective use of the limited resources available. This means in practice that LEDA staff often works together with the staff of government agencies and NGOs. A LEDA with only two or three agricultural extension workers of its own, can thus easily mobilize additional ones if this is necessary for a certain activity.

Another example is the active involvement of commercial banks in the implementation of credit schemes for small farmers and entrepreneurs, instead of the LEDA itself carrying out credit operations. In one case PRODERE was able to persuade a commercial bank to open a branch office in an area where the LEDA was involved in credit operations and where no bank was present. In another case, the bank which had custody over the LEDA guarantee fund, while financing all credits from its own resources, detached one of its employees to the LEDA so that the LEDA and the bank together could follow on the credit portfolio.

Essentially, LEDAs carry out two types of economic development activities. On the one hand they provide direct support in the areas of entrepreneurship promotion and business development services. On the other hand they provide indirect support by promoting a more favourable climate for the creation and growth of small enterprises. Direct support is provided in such areas as information on technology and markets, basic business management and vocational training, counseling and financial assistance. LEDAs also help identify business opportunities and assist in the preparation and formulation of business plans to facilitate the start-up of small enterprises and cooperatives. They thus provide a comprehensive range of support services and help introduce new management techniques and new technologies. They also help mobilize resources and facilitate access to credit.

In terms of indirect support LEDAs can play an important role in facilitating ongoing processes of local planning and decentralization, promoting sectorial public and private investment in the area through lobbying and marketing campaigns and attracting international resources and investments.

The LEDAs promoted by PRODERE paid particular attention to entrepreneurship promotion to help transform business ideas into entrepreneurial ventures. Entrepreneurship promotion was necessary since most actors in the local economy were subsistence farmers, traditionally excluded from the wider economy. Entrepreneurship promotion prepared them to access credit, training and information about markets and technology which in turn helped them improve their productivity and thus become an active participant in the market economy. Awareness raising was fundamental in this context. Through training and information potential entrepreneurs became acquainted with such basic concepts as: competition, commitment, willingness and ability to take risks, initiative, creativity and accountability.

Credit

Since in many of the areas covered by the LEDAs commercial credit was (hardly) available PRODERE also dedicated considerable efforts to facilitating access to credit and introducing new credit mechanisms as appropriate. The credit programmes of the LEDAs promoted by PRODERE provided small farmers and entrepreneurs an opportunity to access credit at market rates and establish a credit record that would eventually enable them to become regular customers of commercial financial institutions. In PRODERE resources amounting to some $17.7 million were channelled through the LEDAs it helped to establish. LEDA staff were trained in screening loan applications by checking whether they were considered good credit risks by their peers. Once an applicant had passed this screening, the LEDA assisted the applicant in preparing a loan application and business plan. Between September 1991 and June 1995 some $15.2 million in credits was actually disbursed to 334,062 direct beneficiaries. More than a quarter of the beneficiaries were women. A little over half of the loans were short-term, mostly for basic staple production and micro-enterprises. The rest was for longer term projects such as livestock raising, coffee production, crop diversification and agro-industry.

The LEDAs used different lending mechanisms to channel PRODERE credit funds. These include loan guarantees, direct lending through administrative agreements with established credit institutions such as banks and credit unions, seed banks, revolving loan funds, materials, usufruct, bridging operations, co-financing and credit lines. The use of these different mechanisms enabled the LEDAs to both meet the diverse credit needs of their constituents and use its funds as leverage with existing financial institutions.

The different credit programmes all shared a set of common goals. First of all, they sought to expand lending to help satisfy the unmet demand for credit in the areas covered by the LEDAs. In Honduras and El Salvador for example, it was not necessary to set up new credit mechanisms in all of the municipalities served by the LEDAs since in some case existing credit programmes could meet the needs of PRODERE beneficiaries. Secondly, they strived at increasing the effectiveness of the programmes by increasing their economic impact and reducing loan losses. Thirdly, they aimed at increasing the profitability of the borrowers, which should help expand employment opportunities. Lastly, they aim at providing a source of income for the LEDAs from the administration of their loan portfolio to cover the cost of providing business development services to farmers and local businesses.

The credit activities undertaken by PRODERE produced a wide range of benefits. Not only were small farmers able to improve the diets of their families by increasing production, but more importantly, individual farmers and groups of farmers used the loans to introduce non-traditional crops and invest in processing technologies that helped improve added value. The introduction of such new technologies could eventually transform the production system of the areas covered and bring about a qualitative improvement in living standards.

The credit funds made available through PRODERE also enabled the LEDAs to establish themselves as serious partners for development efforts funded by others. Thus, during and after PRODERE, the LEDAs were able to mobilize resources for their activities from, among others, Canada, the European Union, Germany (GTZ), the Netherlands, USAID (incl. Peace Corps), a number of UN funds and agencies and a wide range of NGOs. Additional funds fro credit activities were provided by the IDB's Multilateral Investment Fund, the Central American Bank for Economic Integration and the World Bank. In the case of El Salvador the LEDAs became partners in financial institutions created especially with the aim of providing credit to small and medium-sized enterprises.

The table below summarizes the overall size and impact of PRODERE credit activities, while the examples that follow the table describe in more detail the credit operations of individual LEDAs.

The case of the tinsmith in Ocotepeque shows the impact that a LEDA loan can have on an individual and the community. When the LEDA in Ocotepeque had its first contact with this borrower, he was making small items such as buckets, bread pans, and gutters from his farm several hundred meters from the nearest road. The LEDA approved a loan to finance his participation in a course on manufacturing grain silos for use by small-scale farmers. He then invested another portion of the loan in the construction of a new home and workshop in a more visible and accessible location and used the rest to buy enough high quality galvanized metal to respond to the growing demand for household silos. By moving closer to his customers and diversifying his product line, the tinsmith greatly improved sales volume and profits. He then received a second loan to buy more materials to allow him to keep up with orders. The credit not only had a positive impact on the lives of the tinsmith and his family, but also benefited families in the entire area of Ocotepeque, who can now buy low-cost, durable, galvanized metal silos to store their crops, keeping them fresh for later consumption or for sale during periods when the prices for their crops are at their peak.

In Ocotepeque, which was home to 14,000 Salvadoran refugees from 1980 to 1992, the LEDA made close to 1,400 loans. The loans ranged in size from $100 to $2,000, averaging $533. At the end of June 1994, the $736,200 in disbursed credits had resulted in the creation of 3,600 jobs. Ninety percent of the loans were for agricultural production, particularly coffee and basic food crops. Six per cent was for fish farming and four per cent for small agro-industrial, manufacturing, or service businesses. Many agricultural loans were made to groups of farmers who pooled their resources to lease large plots of land, which they farmed collectively. These farms produce mostly cash crops and some subsistence crops to supplement those produced on the small plot of land owned by each member of the group. So far, the default rate is below 10% and no loan funds have had to be written off. This low default rate is due in part to the LEDA's creative and effective loan policies that incorporate both character and peer lending techniques.

In Ixcan, the LEDA lent to farmers' cooperatives, peasant associations and small entrepreneurs. At the end of June 1994, the Ixcan LEDA had approved almost $400,000 in loans, varying greatly in size depending on their purpose. For example, the average size of loans for the cooperative production of organic coffee was $17,765, while loans to small-scale coffee farmers averaged $178. These loans generated over 1,000 new jobs. All credit beneficiaries received training from the LEDA in basic business management as part of an effort to increase the success rate of the ventures. The loan repayment rate to date is 95%.

In Ixil, the LEDA focussed on diversifying the agricultural sector by introducing profitable cold-weather crops that withstand an eight-hour trip over bumpy roads to the market in Guatemala City. It also approved a loan of $254,465 to the Association Chajulense to help finance the production of organic coffee. The credit was used for crop maintenance, drying and storage equipment and facilities. The directly benefited 2,207 farmers and indirectly benefited another 11,185. The LEDA also supported a number of small enterprises, including two brick and roof tile manufacturers and a carpentry shop, with credits of between $1,000 and $2,000. At the end of June 1994 the LEDA had approved slightly over $700,000 in loans. These loans had an average size of almost $90 and were used mainly for coffee production, bee-keeping, and vegetable crops. Some 85%

of the loans were for periods of three to five years. Given the Ixil Indian's tradition of communal activities, it is not surprising that 53 per cent of the loans were made to groups of farmers.

At the end of June 1994 the LEDA in Chalatenango had approved loans for a total amount of some $1.5 million. These loans helped create roughly 5,500 jobs. Half of the loans were to women, as compared to an overall PRODERE average of around only 10 per cent. The high proportion of loans to women in El Salvador reflects the large number of women-headed households after the civil war that killed so many young men. The default rate on these loans is currently below 10 percent.

In Morazan, which experienced some of the heaviest fighting during the conflict, a total of $2.8 million in loans was disbursed to area farmers and small entrepreneurs. These loans covered several sectors including traditional agriculture, livestock raising, diversified crop production, service, manufacturing and trade. By far the largest investment was in the area of trade. Most loans were made to individuals. The loans, which averaged $350 each, created roughly 10,000 jobs. The Morazan LEDA targeted specific sectors of the local economy in an effort to encourage the formation of peer groups of small entrepreneurs engaged in similar businesses, facilitating joint purchasing and marketing.

In Nueva Segovia farmers traditionally had a hard time getting loans from financial institutions. In part, because so many did not hold clear title to their land. Farmers either borrowed from informal sources at interest rates of more than 200% a year or had to sell their crop to intermediaries at below market prices before the planting season to obtain the cash required to buy seeds and other necessary inputs. To overcome this situation, the LEDA set up a network of savings and loans associations to serve the needs of farmers and small entrepreneurs in Nueva Segovia.

4

LESSONS LEARNED

From the outset PRODERE aimed at ensuring the sustainability of the LEDAs it helped establish. The experience of the programme has shown that to achieve sustainability attention needs to be paid to social, political, technical and financial aspects.

SOCIAL SUSTAINABILITY

To achieve social acceptance public awareness raising turned out to be essential. The regions where the LEDAs were introduced were not only the least developed economically but had also been the most affected by violent conflict. As a result PRODERE needed to gain acceptance and achieve a change of mentality before even being able to introduce the local economic development concept. The small farmers and entrepreneurs who were the direct beneficiaries of the programme as well as local policy makers had to be made aware that in the post-conflict situation they themselves were responsible for getting the local economy on track. In this context it was necessary to create faith in their own capacities and to demonstrate that by using their own local resources more effectively they could attract outside resources without waiting for support from the central government.

The programme also had to work on policymakers at the central government level, especially since it was operating in politically conflictive areas. At the national level, decision makers had to understand that the activities carried out at the local level were by no means subversive but rather were in line with national policy and that some of the local level initiatives could become elements of national policies.

The key to ensuring social acceptance and sustainability of the LEDAs was the identification of needs by the local population. Although a time-consuming process it was essential to ensure local "ownership" of the LEDA and arrive at a consensus on priorities. It is only in this way that the

LEDA will be perceived as a legitimate organization and receive the necessary support as witnessed by growth in membership, active participation by members in LEDA activities and the effective provision of services.

TECHNICAL SUSTAINABILITY

Technical sustainability refers to the capacity of the LEDA staff and member organizations to handle effectively most of the day-to-day services the LEDA provides. The key to technical sustainability of the LEDAs has been local capacity building. To this end most LEDAs set up technical committees to advise and guide LEDA staff. Training LEDA staff in technical areas, instead of relying on outside organizations for the provision of services, contributed significantly to the independence of the LEDAs. In addition, the fact that most services were provided by local staff was much appreciated by the members of the local community. Often, local staff is able to solve technical matters in a more appropriate (and understandable) way than external consultants, who do not necessarily speak the local language and may be unfamiliar with local customs.

Technical sustainability was also strengthened through a process of networking with similar institutions inside and outside Latin America. For instance, it was only after a study tour to Europe in 1992 that the methodology and the principles of local economic development were fully accepted and became the basis for PRODERE's later interventions. In this respect it is significant to note that even after the end of the programme the network of LEDAs in Central America continued to function and at the time of writing this paper had just set up its own website (address), including a detailed profile of each LEDA (see also the Annex to this paper).

FINANCIAL SUSTAINABILITY

Financial sustainability was a fundamental objective of the programme since it was the only way to ensure that the LEDAs would continue to function after the end of the programme. LEDAs, given their membership structure, their participatory character and their comprehensive range of services, are relatively expensive to operate. Even in industrialized countries similar agencies often receive considerable external subsidies since it is not realistic to expect that LEDA members and clients can finance all operations. The PRODERE experience has shown that during a period of at least two to three years external support is necessary to guarantee a minimum of services and work towards full financial sustainability.

The main source of income for the LEDAs in the case of PRODERE was through their participation in the credit activities, particularly guarantee funds. The administrative income derived from the LEDA's involvement in credit operations is usually sufficient to finance a number of basic services. LEDAs can generate additional income by charging membership fees, becoming executing agency for international and national technical cooperation projects, charging for services (incl. training) provided and mobilizing external resources. In this latter respect it is important to note that after the end of the programme LEDAs in Central America have received support from a variety of agencies and organizations, including the Inter-American Development Bank's Multilateral Fund for Investment, the Central-American Bank for Economic Integration and the German Agency for Technical Cooperation (GTZ). In fact, the income the LEDAs have received through such partnerships has exceeded the amount involved in the support they received from PRODERE. It is also worth noting that the LEDAs in El Salvador have become partners in a new financial institution specializing in micro-credits.

As mentioned in the paper, one of the reasons that LEDAs entered into credit operations was the fact that in many of the areas in which they operated there were practically no commercial credit facilities available to the farmers and small entrepreneurs. The PRODERE approach to credit operations was based on the following principles:

- services had to available in close proximity to the clients;
- character references played an important role in the appraisal of credit applications;
- collateral requirements were kept flexible;
- documentation and repayment modalities were kept as simple as possible;
- credit applications were processed and approved locally.

REFERENCES

Arias Sanchez, O.; *El PRODERE: Una cooperación novedosa y efectiva en apoyo a la paz y al desarrollo en centroamerica* (Hombre de Maiz n°31, pp. 40-42, March 1995)

Italian Cooperation; *PRODERE, Le strategie, i metodi ed i resultati di un Programma per lo Svilupo Umano, la Pace e la Democrazia in America Centrale* (1996)

Danieri, F. and Mancinelli, G.; *Situational Analysis of the system: LEDA* (ILO/UNOPS/Red Centroamericana de ADELs/AIESEC, 1997)

Fundación Arias para la Paz y el Progreso Humano (Coordinador), Refugee Policy Group, Friedich Ebert Stiftung, Centro Internacional para el Desarrollo Económico, OECD; *PRODERE, Informe de Evaluación Externa* (1996)

Gambo, N., Diaz, M. and others; *Especial DESARROLLO HUMANO: PRODERE, un modelo de cooperación al desarrollo* (Hombre de Maiz n°36,pp. 6-52 August-September 1995)

Gibbons, C., Conway, M., O'REGAN, F.; *Regional Strategies for Employment and Poverty Alleviation. Domestic discussions and international models* (The ASPEN Institute, 1994)

Lazarte, A., Paredes, P.; *PRODERE: Central American Programme carried out by UNDP/UNOPS, ILO, UNHCR, WHO/PAHO and other UN organizations* (In : Building a Consensus on International Cooperation for Social Development. World Summit on Social Development.pp. 11-14 Copenhaguen, 1995)

Lazarte, A., Cruz, R.; *La experiencia de credito en PRODERE: Informe Final al 31 de Julio de 1995* (1996)

Martens, J., van Boekel, G.; *ILO Local Economic Actions in Central America: the case of PRODERE* in: The design and implementation of strategies for local employment and economic development (ILO/Commission of the European Communities, 1993)

PRODERE, Coordinación Regional; *Informe Final* (1996)

PRODERE Edinfodoc; *PRODERE, Programa de Desarrollo para Desplazados, Refugiados y Repatriados en Centroamérica* (1993)

Red Centroamericana de ADELS; *Agencias de Desarrollo Económico Local. Una red centroamericana en camino* (1995)

Revilla, V.; *Examen Crítico de la participación de la OIT en el Programa PRODERE* (ILO, 1994)

SICA (Sistema de Integración Social/Comisión Regional de Asuntos Sociales; *Propuesta Centroamericana: de Esquipulas al Desarrollo Social Sostenible. Cumbre Social sobre Desarrollo social. Copenhaguen* (1995)

UNOPS; *Informe de la Reunión Tripartita Regional Final del PRODERE de Julio 1995* (UNOPS, 1996)

5

ECONOMIC ASPECTS OF SUSTAINABLE DEVELOPMENT IN INDIA

INTERNATIONAL COOPERATION

Decision-Making: Coordinating Bodies

Almost all ministries of the Government of India are involved in decision making for sustainable development. However, major participation is by the Ministries of External Affairs, Environment and Forests, Agriculture, Water Resources, Finance, Industries, Rural Development, Commerce, Non Conventional Energy Sources, Finance and the Planning Commission.

Coordination within the different bodies of the Government in India is mainly through consultative meetings and discussions. There are inter-ministerial and inter-departmental committees, Core Groups for coordination to formulate the optimum policy and legislation on issues concerning international cooperation/development assistance for Sustainable Development.

The authority for decision-making lies with the ministries of the Government of India .In India the decentralization of powers and decision making has been done at the grass root level. The authority for decision making has been delegated to provincial/State level in several areas with the exception of core areas of national interest. In matters of international relations and co-operation the central government coordinates the overall decision making process.

Decision-Making: Legislation and Regulations

Government of India has have formulated legislation, regulations and policy instruments to address matters concerning cooperation for Sustainable Development at sub-regional, regional and International level. There are legislation regulations and policy instruments framed to fulfill obligation under the agreements signed under the International Conferences, MEAs, etc. They are:-

INDIAN ENVIRONMENTAL LEGISLATIONS

- The Environment (Protection) Act, 1986
- The Water (Prevention and Control of Pollution) Act, i974, as amended up to 1988
- The Water (Prevention and Control of Pollution) Cess Act 1977, as amended by Amendment Act, 1991
- The Air (Prevention and Control of Pollution) Act, 1981, as amended by Amendment Act, 1987
- National Forest Policy, 1988
- Forest (Conservation) Act, 1980
- The National Environment Tribunal Act, 1995
- The Public Liability Insurance Act, 1991
- Re-cycled Plastics Manufacture and Usage Rules, 1999
- Manufacture, Use, Import, Export and Storage of Hazardous Micro-Organisms
- Genetically Engineered Organisms or Cells rules, 1989
- Hazardous Wastes (Management and Handling) Rules, 1989
- Bio-Medical Waste (Management and Handling) Rules, 1998
- Municipal Solid Wastes (Management & Handling) Rules, 2000
- Noise Pollution (Regulation and Control) Rules, 2000
- Ozone Depleting Substances (Regulation) Rules, 2000
- New Biodiversity Bill - 2000

- The Prevention and Control of Pollution (Uniform Consent Procedure) Rules, 1999

Selected Policy initiatives taken by the Ministry of Environment and Forests towards sustainable development

- National Environmental Action Plan for Control of Pollution
- Urban Pollution
- Vehicular Pollution
- Environmental Epidemiological Studies
- Environmental Management System (EMS)
- Uniform Consent Procedure
- City Afforestation Programme for Mitigating Pollution
- National Biodiversity Strategy and Action Plan (NBSAP)
- Biosafety Protocol
- Rules on the management of Lead acid Batteries
- Regional Development Strategy based on Carrying Capacity concept
- Development of Management Tools for preventing environmental degradation
- Establishment of Indian Centre for Promotion of Cleaner Technologies (ICPC)
- Water Quality Standards for Sewage
- Technology for Sewage Treatment
- Water Conservation through recycling
- Joint Forest Management
- National Forest Action Programme
- National Forestry Research Plan

AUTHORITIES UNDER ENVIRONMENT PROTECTION ACT, 1986

A National Environmental Appellate Authority has been constituted to hear appeals with respect to rejection of proposals from the environmental angle. The objective is to bring in transparency in the process and accountability, and to ensure the smooth and expeditious implementation of developmental schemes and projects.

An Environmental Impact Assessment Authority for the National Capital Region has been constituted to deal with environmental protection problems arising out of projects planned in the National Capital Region (NCR). An Aquaculture Authority has been constituted to deal with the situation created by the Shrimp Culture industry in the coastal States and Union Territories. The Central Ground Water Authority for regulation and control of ground water management has initiated action regarding registrations for ground water pollution/depletion. It has also initiated a mass awareness programme. Besides this, different authorities have been created for dealing with specific problems in the States of Tamil Nadu and Maharashtra.

The trade policy components of the Indian reform process undertaken since July 1991 have been motivated by a recognition of the important role that trade can play in promoting sustained economic growth in the context of sustainable development. The expanded scope for specializing in areas of comparative advantage is manifest in the improved growth performance of the economy. Furthermore, while exports have responded to the removal of the anti-export bias of a protectionist environment, domestic industry appears to have been stimulated by the expanded availability of imports and capital goods, and the challenge of competing in the international market place. The positive response of Indian industry to deregulation is amply demonstrated by the capital goods sector. The capital goods industry, which witnessed negative growth of 12.8% in 1991-92, registered an average growth of about 23% during 1994-96.

In an effort to remove the anti-export bias of existent policies, improve the efficiency of resource allocation as well as the competitiveness of domestic markets, India has made steady progress in eliminating quantitative restrictions, licensing, and discretionary controls over imports since 1991. Imports of capital goods, raw materials, and components have been de-licensed, tariffs on such imports have been reduced substantially, and tariff categories have been streamlined and simplified. As a result, all goods can now be freely imported and exported, except those belonging to two negative lists.

Indian companies have made investments in several countries all over the world including the neighboring countries in the Indian sub continent.

Decision-Making: Strategies, Policies and Plans

The Ministry of Environment and Forests functions as a nodal agency for United Nations Environment Programme (UNEP), South Asia Co-operation Environment Programme (SACEP), and International Centre for Integrated Mountain and Development (ICIMOD), International Union for Conservation of Nature and Natural Resources (IUCN) and various international agencies, regional bodies and multilateral institutions. India is signatory to the following important international treaties/ agreements in the field of environment: (i) International Convention for the regulation of Whaling; (ii) International Plant Protection Convention; (iii) The Antarctic Treaty; (iv) Convention on Wetlands of international importance; (v) Convention on International trade in Endangered Species of Wild Flora and Fauna; (vi) Protocol of 1978 relating to the international convention for the prevention of pollution from ships; (vii) Vienna Convention for the protection of the Ozone Layer; (viii) Convention on Migratory Species; (ix) Basel Convention on Trans-boundary movement of hazardous substances; (x) Framework Convention on Climate Change; (xi) Convention on conservation of bio-diversity; (xii) Montreal Protocol on the substances that deplete the ozone layer and; (xiii) International Convention for Combating Desertification.

The Ministry and its agencies cooperate with various countries such as Sweden, Netherlands, Norway, Denmark, Australia, U.K., U.S.A., Canada, Japan, Germany among others bilaterally and from several UN and other multilateral agencies such as the UNDP, World Bank, Asian Development Bank, OECF (Japan) and ODA (U.K.) for various environmental and forestry projects.

India believes that environmentally harmful processes should be stopped and that over-exploitation of non-renewable resources should be controlled. However, the specific production process to be used would depend upon the absorptive capacities and development priorities of the country concerned and hence, no global harmonized standard for production processes can be developed. The solution lies not in unilaterally banning trade, but rather in transferring technology and offering prices to developing countries for commodities, which would not then necessitate their overexploitation or jeopardize their development priorities.

India is signatory to various regional and international agreements, which provide financial assistance for development cooperation.

The issue of technology transfer from developed to developing countries has been a recurring theme in all multi-lateral and bilateral negotiations/ discussions and is at the heart of the North-South divide. At the same time, efforts to develop a mechanism to actually to carry out such a transfer have met with only moderate success so far. Indian research and development (R&D) efforts, unlike in the developed countries, has remained largely in the Government domain e.g. the CSIR and University laboratories. In India, initial steps have been taken through the National Productivity Council established with GEF assistance and UNEP as the implementing agency. The initiative however, has a limited scope in so far as it only addresses a small portion of the problem.

In the Export-Import Policy of 1992-97 conscious efforts were made to dismantle various protectionist and regulatory policies and accelerate the country's transition towards a global economy. The 8th Five Year Plan (1997-2002) seeks to consolidate the gains of previous policy and further carry forward the process of liberalization involving further deregulation, simplification of procedure and removal of QRs. Efforts have been made to mobilize resources from domestic and external sources for development, cooperation in sustainable development including environment protection.

Decision-Making: Major Groups Involvement

All the major groups identified in Agenda 21 are involved in decision making in different capacities. Participation of these groups is ensured through consultative meetings and discussions at local, state and national levels.

Government of India has made public hearings mandatory for developmental projects wherein affected person, stakeholders are given opportunity of hearing/ discussion before arriving at a decision. Public participation is also an important step in every major decision for social, economic and sustainable development. Participation is encouraged by bringing in transparency in decision making.

Major groups which participate in international cooperation activities programmes are indigenous groups, NGOs, Industrial Associates, Investigators, Research Institutions, Advocates, etc. The Government facilitate the participation of various groups in arriving at a decision in a more participatory manner.

India has had modest, but increasing success in attracting private capital flows. Furthermore, much of these private capital inflows into India have been of the non-debt creating variety, which has helped boost the balance of payments as well as the availability of invertible resources in the economy. The international community is very positive about India's effort to achieve a high rate of growth. After the advent of liberalization which was initiated in 1991, the involvement of private sector (local and foreign) has been encouraged.

The bulk of India's population is still rural and engaged in agriculture which generates nearly one third of India's national income. While economic reforms have been mainly confined to the industrial sector, they have affected agriculture since these reforms have significantly altered relative prices and protection. A fall in agricultural prices has an undesirable effect on the welfare of the people who depend upon agriculture. In India's context globalization in terms of liberalization of agricultural trade can have a profound impact on the poor. India feels that there remains a need for an alternative agricultural trade agenda that promotes greater food-self sufficiency and food security in the developing countries rather than promoting global harmonization of standards in subsidies.

The most adversely affected groups are thus small farmers and small scale industries. These groups are protected by provision food security, better prices for their products, incentives and other facilities to improve their performance to make them competitive.

Programmes and Projects

Several activities and programmes involving multilateral financing are ongoing in India which include

Global Environment Facility through the World Bank, UNDP and UNEP: India is the second largest recipient of GEF funding. The salient feature of the GEF portfolio are: a diverse and varied portfolio comprising projects that are environmentally, socially and financially sustainable; projects involving a range of issues and approaches to address the questions of innovation, experimentation, demonstration, cost effectiveness and replicability; projects that are country-driven, based on national priorities; capacity building, human resources and skills at the community level and into Government.

The Country cooperation Framework- I Environment Programme through the UNDP: Development Objective: The thrust areas reflect the national policy and plan statements

– (i) management of natural resources (ii) capacity building for decision making (iii) management of development (iv) information, advocacy and participation.

Montreal protocol: The Protocol sets out a time schedule for freeze and reduction of ODS or controlled substances. A Multilateral Fund was established by the parties to assist developing countries meet the control measures as specified in the Protocol. It assists the Government and the industry to design, implement, monitor and evaluate ODS phase-out projects and programmes in the aerosols/foam/solvent refrigeration and fire extinguishing sectors, covering large, medium and small scale enterprises. The MOEF is the national executing agency for the Institutional Strengthening projects for the phase-out of ODSs under the Montreal Protocol. In Asia, India is number three in receiving funds for CFC phase out programme, next to China and Malaysia.

Capacity 21 Initiative: There is only one Capacity 21 project in India which is being implemented by the Indira Gandhi Institute for Development Research (IGIDR) through the Ministry of Environment & Forests. The main objective of the project is to build capacity at various levels of Government, national institutes and the community at large through NGOs by introducing concepts of environmental economics into their resource use and planning decisions. Specific interventions of natural resource accounting through practical applications at policy and field levels include – Air quality, Water Quality, Biodiversity and Common Property Resources. IGIDR have come out with documentation on the above areas.

LIFE programme of UNDP: The Programme of Action for Sustainable Development Worldwide, Agenda 21, was adopted by more than 178 governments at the Earth Summit in Rio de Janeiro in 1992. The Local Initiative Facility for Urban Environment (LIFE) was launched by UNDP at this Summit. The main goal of the programme is to help city dwellers to help themselves, to find local solutions to local problems.

SDNP: The Sustainable Development Network Programme is a UNDP initiative launched globally in 1990 to make relevant information on sustainable development readily available to decision-makers responsible for planning sustainable development strategies.

In India there are several ongoing projects which are being implemented through various bilateral programmes. Some of these include CIDA, IDRC, OECF/Japan, JICA, and other bilateral cooperation programmes with countries inter -alia including U.K., Norway, Sweden and Germany. The main thrust of these programmes is on

- Basic human needs
- Women in development
- Support to infrastructure
- Private sector development
- Environment
- Good governance
- Developing Eco-Friendly goods and technologies

The largest share is for poverty eradication, natural resource protection and capacity building in that order. The amounts are miniscule compared to the needs of the country.

- Environment Management Capacity Building Project (World Bank) (UNCTAD Project)
- Strengthening capacities for trade and environmental policy integration in India and trade environment investment (UNCTAD)
- Strengthening Research and Policy making capacity on trade and environment in development countries (UNCTAD).

India is part of SAPTA (South Asian Preferential Trade Agreement) and BIMSTEC and in these regional groupings, the question of market accessibility and trade has received due consideration.

Status

The balance continues to weigh in favor of ODA and bilateral assistance. Though there has been a slight increase in private flows.

Challenges

Some of the major challenges for building partnership with countries which are in various stages of economic development, are inadequate implementation of commitments, transfer of technology, financial constraints market access and standards.

The programme areas/issues of Agenda 21 which require most immediate attention for bilateral/multi-lateral cooperation are fulfillment of obligations of transfer of technology, financial assistance, capacity building, public participation, involvement of NGOs and private sector, R&D institutions and scientific/business community.

Some of the major challenges in building partnerships with NGOs and private sector scientific community.

- recognizing the sustainable development as mutual goal
- lack of understanding of issues for achieving sustainable development
- development of various tools/instruments and their implementation
- paucity of financial resources.

There are DGFT, Inter-Ministerial Committees, Core Groups, Bureau of Indian Standards which are responsible for various types of standards setting. The Committees/ agencies coordinate which each other and also interact with international agencies for avoidance of technical barrier in the flow of trade. However, some with the consideration and information gap pose challenges.

Capacity-building, Education, Training and Awareness-raising

The following initiatives have been taken to promote public awareness on environmental issues in general they include awareness about international co-operation and sustainable development

Environmental Information systems

Since environment for sustainable development is a broad-ranging, multi-disciplinary subject, a comprehensive information system on environment has necessarily involved effective participation of concerned institutions/ organisations in the country that are actively engaged in work relating to different subject areas of environment. Realizing the importance of Environmental Information, the Government of India, in December 1982, established an Environmental Information System (ENVIS) as a plan programme. The focus of ENVIS since inception has been on providing environmental information to decision makers, policy planners, scientists and engineers, research workers, etc. all over the country. A large number of nodes, known as ENVIS Centres, have been established in the network to cover the broad subject areas of environment. Similarly the Sustainable Development Networking Programme (SDNP) is a UNDP/ IDRC initiative launched world-wide in 1990 to make relevant information on sustainable development readily available to decision-makers responsible for planning sustainable development strategies. SDNP-India is being implemented by the ENVIS a GOI programme, over a period of three years.

Environment Education and Awareness Generation

Capacity building initiatives by the GOI in various sectors relating to the environment are an ongoing process and form an integral part of most projects and programs on sustainable development. Steps are taken to involve NGOs in organizing orientation training courses for teachers. The two centres of Excellence, namely the Centre for Environment Education, Ahmedabad and the CPR Environment Education Centre, Chennai provide the backup support to the NGOs. GOI's efforts towards Non-formal Environmental Education & Awareness include:

- National Environment Awareness Campaign (NEAC)
- Eco-clubs

- Paryavaran Vahinis
- Seminars/Symposia/Conferences/Workshops
- Publicity through State Transport Bus Panels
- Films on Environment related Areas
- Communication & Awareness Programme of NAEB
- Early Childhood Education Scheme started in 1982 to reduce the drop out rates and to improve the rate of retention of children in primary school in educationally backward areas of the States of Andhra Pradesh, Assam, Bihar, Jammu & Kashmir, Madhya Pradesh, Orissa, Rajasthan, Tripura and West Bengal.

None of the government and civil society actions would mean much without the media. The GOI's Ministry of Information & Broadcasting, through the mass communication media consisting of radio, television, films, the press, publications, advertising and traditional mode of dance and drama plays a significant part in helping the people to have access to free flow of information on issues including environmental protection.

Programmes (as mentioned above) have been carried out to nurture technical experts and professionals in international relations through UNCTAD.

The areas of project formulations, project management and implementation require strengthening. Technical experts, managers and administrators are coordinating their efforts to ensure co-operation is relevant and practical.

Research and Technologies

India has increasingly recognized the critical role of technology. India has a strong base in technology and R&D institutions. As regards Environmentally Sound Technology, India still needs technical, financial assistance. These issues are the current priorities in the programmes/policies being implemented in achieving sustainable development.

Public investment for sustainable development through fiscal incentives and concessions has always been emphasized. Since energy-efficient technologies and non-conventional energy technologies directly improve the protection level of the atmosphere, several tax concessions, 100% depreciation allowance, and investment subsidies have been made widely available.

The promotion of environmentally sound technologies through international cooperation is mainly in the form of FDI Joint Venture. However, ESTs as envisaged in various MEAs are not being transferred to developing countries on fair and favorable terms and conditions.

Public investment for sustainable development by providing fiscal incentives and concessions has been emphasized. Since energy-efficient technologies and non-conventional energy technologies directly improve the protection level of the atmosphere, several tax concessions, 100% depreciation allowance, and investment subsidies have been made widely available.

Financing

The goal of Agenda 21 was in part to raise additional external funds for sustainable development activities by increasing bilateral and multilateral Official Development Assistance (ODA) to 0.7% of GNP from donor countries. The fact remains that many of the developing countries are experiencing a net outflow of resources. The average ODA in the post-Rio period 1993-95 has been lower than in the period 1990-92, both in absolute terms and as a percentage of GNP. In fact, ODA at an average of 0.29% of GNP in the 1993-95 period has been the lowest in decades.

In the context of declining ODA, we have to find adequate financing for environmental measures either from our own budgetary resources or by generating funds from the private sector. Domestic resources will continue to be an important source for financing sustainable development and countries need to develop an enabling environment to encourage the mobilization of additional financial resources. Key elements include a sound macroeconomic framework, a dynamic private sector, governance and participatory mechanisms. Special attention is being given to fiscal and budgetary policies, tax collection and transparency.

The following sources are being tapped for financial assistance and there are:

- Bilateral sources other than ODA;
- Private sources (specify forms: e.g. foreign direct investment, joint ventures, etc.); and
- Multilateral sources.

Cooperation

The Ministry of Environment and Forests functions as a nodal agency for United Nations Environment Programme (UNEP), South Asia Co-operation Environment Programme (SACEP), and International Centre for Integrated Mountain and Development (ICIMOD), International Union for Conservation of Nature and Natural Resources (IUCN) and various international agencies, regional bodies and multilateral institutions.

India is signatory to the following important international treaties/ agreements in the field of environment:

(i) International Convention for the regulation of Whaling;

(ii) International Plant Protection Convention;

(iii) The Antarctic Treaty;

(iv) Convention on Wetlands of international importance;

(v) Convention on International trade in Endangered Species of Wild Flora and Fauna;

(vi) Protocol of 1978 relating to the international convention for the prevention of pollution from ships;

(vii) Vienna Convention for the protection of the Ozone Layer;

(viii) Convention on Migratory Species;

(ix) Basel Convention on Trans-boundary movement of hazardous substances;

(x) Framework Convention on Climate Change;

(xi) Convention on conservation of bio-diversity;

(xii) Montreal Protocol on the substances that deplete the ozone layer and;

(xiii) International Convention for Combating Desertification.

The Ministry and its agencies cooperate with various countries such as Sweden, Netherlands, Norway, Denmark, Australia, U.K., U.S.A., Canada, Japan, Germany among others bilaterally and from several UN and other multilateral agencies such as the UNDP, World Bank, Asian Development Bank, OECF (Japan) and ODA (U.K.) for various environmental and forestry projects.

Sustainable development is an important consideration in Bilateral trade agreements that India has signed.

The role can be described as active. India has been the spokesman of the G-77 and China on Climate Change and has played a major role in UNEP.

TRADE

Decision-Making: Legislation and Regulations

India is among the countries, which are in the vanguard of environmental protection. India has environmental standards for products and processes, has environmental impact assessment and has introduced environmental audit as well as an eco-labeling scheme. India believes that environmentally harmful processes should be stopped and that over-exploitation of non-renewable resources should be controlled. However, the specific production process to be used would depend upon the absorptive capacities and development priorities of the country concerned and hence, no global harmonized standard for production process can be developed. The solution lies not in unilaterally banning trade, but rather in transferring technology and offering prices to developing countries for commodities, which would not then necessitate their overexploitation or jeopardize their development priorities.

In an effort to remove the anti-export bias of existent policies, improve the efficiency of resource allocation as well as the competitiveness of domestic markets, India has made steady progress in eliminating quantitative restrictions, licensing, and discretionary controls over imports since 1991. Imports of capital goods, raw materials, and components have been de-licensed, tariffs on such imports have been reduced substantially, and tariff categories have been streamlined and simplified. As a result, all goods can now be freely imported and exported, except those belonging to two negative lists. In accordance with the provisions of WTO, a country is required to lift quantitative restrictions on imports imposed for Balance of Payments reasons when the position improves. As the position of foreign exchange reserves is comfortable, India has decided to phase out quantitative restrictions in respect of all items where such restrictions were maintained for balance of payments purposes.

Decision-Making: Strategies, Policies and Plans

With the objective of accelerating the pace of reforms, sustaining high export growth, and enhancing the opportunities for the domestic economy's participation in the dynamics of foreign trade, the Export-Import (EXIM) Policy 1992-97 has been reviewed and revised in several ways during the current year to further phase out quantitative and qualitative restrictions. The revisions include measures for trade promotion, as well as further simplification of procedures. The trade policy components of the Indian reform process undertaken since July 1991 have been motivated by a full recognition of the important role that trade can play in promoting sustained economic growth in the context of sustainable development. Restrictive trade barriers and practices must be curtailed, and tariffs, particularly peak tariffs, on exports of products and services from developing countries reduced so that the benefits of global economic growth are equitably distributed among all countries. Greater trading opportunities can enable developing countries to invest more in environmental protection.

Decision-Making: Major Groups Involvement

It need not be emphasized that poverty is intrinsically related to both trade and investment. In the last World Bank meeting in December 1999 at Seattle, due to lack of understanding of the need of the poverty stricken population by the planners the meeting could not progress further. At the international level linkage between trade and poverty is addressed in India's contribution to the discussions in the CTE of WTO regularly.

Programmes and Projects

The Environmental Impact Assessment programme of the Ministry of Environment and Forests, Government of India is intended to identify "hot spots" in the integration of trade and environment issues.

Status

The expanded scope for specializing in areas of comparative advantage is manifest in the improved growth performance of the economy. Furthermore, while exports have responded to the removal of the anti-export bias of a protectionist environment, domestic industry appears to have been stimulated by the expanded availability of imports and capital goods, and the challenge of competing in the international market place. The positive response of Indian industry to deregulation is amply demonstrated by the capital goods sector. The capital goods industry, which witnessed a negative growth of 12.8% in 1991-92, registered an average growth of about 23% during 1994-96. The main agricultural commodities exported from India are Rice, Tobacco, Spices, Cashew, Oil meals, Fruits and Vegetables fresh and processed, marine products, tea and coffee. As regards imports, the main items are pulses, unprocessed Cashewnuts, vegetable oils and sugar.

Export, as a percentage of GDP at constant prices is less than 1% in India. Hence, local or national environmental problems in India are more associated with domestic production. Trade, as a percentage of GDP, has more or less remained constant at about 20% in India.

Challenges

The linkage between trade and poverty from the environmental point of view is addressed in a multifunctional manner in India. Environmental requirements in the markets for Indian exports could cause loss of exports. This could be unnecessary especially where there are environmental restrictions based on processes and production methods or based on the precautionary principle and where the measures are susceptible to disguised protectionism. This in turn would make less money available to national policy makers to direct towards environmental protection. Poverty, it is said, is the biggest polluter and this holds particularly true for us, as fund constraints may be one of the major handicaps in addressing environmental problems through appropriate national environmental policy. The inability of the developed world to implement provisions of Agenda 21 in letter and spirit has added to the problems.

Capacity-building, Education, Training and Awareness-raising

Information related to trade, investment and economic growth is made available to potential users via the Internet:

Research and Technologies

Financing

India has had modest, but increasing, success in attracting a growing part of private capital flows. Furthermore, much of these private capital inflows into India have been of the non-debt creating variety, which has helped boost the balance of payments as well as the availability of investable resources in the economy. The international community is very positive about India's effort to achieve a high rate of growth.

Cooperation

India believes that in order to make trade and environment mutually supportive, an open multilateral trading system makes possible a more efficient allocation and use of resources. This contributes to increased production and incomes, and lessens the demands on the environment. It also provides the additional resources needed for economic growth and development, and improved environmental protection. Trade measures should be applied for environmental purposes only when they address the root causes of environmental degradation so as not to result in an unjustified restriction on trade. Further, environmental standards valid for developed countries may have unwarranted social and economic cost in developing countries. India believes that global efforts at environmental protection are best addressed through Multilateral Environmental Agreements (MEAs), which contain a package of positive measures, including among them financial and technological transfers and capacity building.

Today, the world stands at a crossroad of history. Five years after Rio, and as we approach the third millennium, it falls on the global community to create a world where there is greater justice and lesser deprivation. Any models based on uneven rewards will not be supported by those members who are not beneficiaries. Credibility and realization of the potential of all international activities can only be achieved through the full participation of all countries in their formulation, implementation, and enjoyment of benefits. India is willing to work with all countries in a constructive manner to realize common goals.

India submits a report on its trade to WTO every four years as part of its trade policy review and reports to IMF and the World Bank.

The World Trade Organization's (WTO) Committee on Trade and Environment has undertaken many of these activities. In its five years of work, it has promoted a dialogue between trade, development, and environment communities and stressed the need for transparency, avoidance of unilateral action to deal with environmental challenges outside the jurisdiction of the importing country, and avoidance of the use of trade restrictions or distortions as a means to offset the difference in costs arising from environmental distortions and protectionism. Further priority work under the Committee and the United Nations Conference on Trade and Development (UNCTAD) should include elaborate studies for better understanding of the relationship between trade and environment, particularly for sustainable development in developing countries. India realizes the vital need for international cooperation - bilateral, multilateral, and regional initiatives - in implementing Agenda 21. India is committed to developing and strengthening the process of international cooperation, which would cover not only cooperation among governments and international agencies but also among other major actors such as the private sector, civil society, and voluntary organizations. The international community should develop the appropriate open, equitable, rule based, cooperative, non-discriminatory, and mutually beneficial economic environment at the international level. It should take into account the special needs of the developing countries, in line with the concept of common but differentiated responsibilities affirmed in Agenda 21. The international community should, therefore, aim to attain the target of 0.7% of GNP for the Official Development Assistance (ODA) from developed countries. There is also an urgent need for new and additional financial resources on a predictable and assured basis from the international community to developing countries. These resources should be available commensurate with the needs and priorities of developing countries and without any conditions.

This information was provided by the government of India to the 5th and 8th Sessions of the United Nations Commission on Sustainable Development. Last Update: February 2000.

CHANGING CONSUMPTION PATTERNS

Decision-Making: Coordinating Bodies

The responsible Government bodies dealing with aspects of sustainable consumption and production patterns include:

the Agricultural Products Export Development Authority (APEDA), New Delhi; Indian Institute of Plantation Management (IIPM), Bangalore; Central Pollution Control Board, Bureau of Indian Standards and National Productivity Council, State Environment Protection Councils, and the National Consumer Council.

Decision-Making: Legislation and Regulations

Legislation which seeks to promote sustainable consumption and production include:

- Environment Protection Act, 1986
- Forest Conservation Act, 1980.

Show Cause notices under Section 5 of the Environment (Protection) Act 1986 have been issued to all defaulting units. In addition, National ambient air quality standards, including ambient noise standards, have been notified. Industries have been directed to instal necessary pollution control equipment within a stipulated time frame.

More stringent norms for vehicular emissions have been notified under the Central Motor Vehicles Rules which came into effect in April, 1996. The supply of unleaded petrol in the four metropolitan cities of Mumbai, Calcutta, Delhi, and Chennai was introduced in April 1, 1995 for use in four wheel vehicles fitted with catalytic converters. The use of unleaded petrol will be gradually extended to other cities in the country.

In order to enhance energy and material efficiency, waste reduction, recycling, public transport and quality of life, norms have been laid down by the Government of India. Industry has adopted, on a voluntary basis, EMS (Environment Management Systems, in order to attain more sustainable production.

Decision-Making: Strategies, Policies and Plans

National Strategy and policies that address the concerns of this area include:

- National Conservation strategy
- Environment Action Programme
- Statement of Abatement of Pollution Control
- National Forest Policy

Specific issues such strategies and policies address are the following:

- increasing energy and material efficiency in production processes;
- reducing wastes from production and promoting recycling;
- promoting use of new and renewable sources of energy
- using environmentally sound technologies for sustainable production;
- reducing wasteful consumption;
- increasing awareness for sustainable consumption.

The ongoing initiatives of Government to improve the environment include preventive as well as promotive measures. Fiscal incentives are provided by the government to encourage the installation of appropriate pollution abatement equipment in the form of customs waivers and soft loans. Industries are encouraged and fiscal incentives support the installation of equipment for pollution control; punitive measures including legal action are taken against defaulting units.

To achieve the goal of pollution abatement, emission and effluent standards for air, water, and noise have been notified. Regular monitoring is carried out and enforcement efforts have been intensified. A m ajority of identified units have already installed the requisite pollution control equipment According to data collected by the Central Pollution Control Board (CPCB) on September 30, 1996, out of 1,551 units belonging to 17 categories of highly polluting industries, 1,259 units have facilities to comply with the environmental standards, 112 were closed, and 180 did not have adequate facilities.

Decision-Making: Major Groups Involvement

At the local and provincial levels, the responsible authorities are the regional offices of the Bureau of Indian Standards (BIS), State Pollution Control Boards (SPCBs), and State Consumer Councils. The view of Major Groups and the public in general are solicited. Standards and criteria are evolved and finalised only after circulating them for public comments and views.

Programmes and Projects

Government Programmes, in partnership with industries, consumer associations and others, to promote sustainable consumption and production patterns include:

- The Ecomark Scheme
- Green Rating
- ISO 14001 Certificate

Twenty-four critically polluted areas in the country have been identified and action plans have been drawn up to improve the quality of the environment in these areas (1997). Adoption of Cleaner Production Technologies and formation of Waste Minimization Circles are being encouraged to minimize environmental pollution. Under the World Bank aided Industrial Pollution Control Project, technical and financial assistance is provided for establishing Common Effluent Treatment Plants (CETPs) in clusters of small scale industrial units. An Eco-Mark scheme has been launched to certify various products of industries which fulfil the prescribed standards of environment-friendly production, packaging, and waste disposal.

The major programmes for new and renewable sources of energy which were developed and enlarged during the Seventh Five Year Plan included National projects on bio-gas development, improved Chulhas, solar, thermal energy utilization, Solar Photo Voltaics (SPV), wind energy, and conversion of bio-mass into energy, energy plantations, and bio-mass gasifiers.

Status

The process of development is sharply raising the consumption of household energy. It is imperative to support the development of non-conventional or renewable sources of energy to sustain the development process. Sun, wind, water, and biomass are renewable, perennial, dependable, and widely available sources of energy. The generation and utilization of energy from renewable sources have tremendous potential. According to available statistical data (information provided in 1997), India accumulates 300 million tonnes of agro residues every year of which only a small quantity is used as direct fuel. The potential of biomass energy is placed at 17,000 MW and of solar energy at SX10l: KWHours/year. Using a conservative assessment, wind power potential in the country is around 20,000 MW and mini hydro-energy 5,000 MW.

The total wave power potential from ocean energy along India's 1600 km coastline is 40,000 MW. Patterns of consumption by the very poor, even when unsustainable in the short term, must be regarded primarily as survival consumption. Overuse of agricultural land, over-grazing of pasture land, and the depletion of forests for fuel wood are all manifestations of a survival economy. To speak of such consumption as being unsustainable, and hence requiring change, without addressing the human condition that leads to such consumption, is not only unethical but also impractical.

The efficient usage of energy, water and other materials by industries and by households, is gaining recognition, acceptance and picking up progressively. Recycling and reuse has long been an established tradition in Indian society. Deposit and refund practices have been quite widespread in the consumer industry in India. An extensive and effective collection and recycling system for wastes such as glass, tin scrap iron, brass, rubber, paper, and plastics thrives in the non-formal sector. Consumers are increasingly aware of the health effects of residual pesticides and fertilizers. Textile, leather, and other industries are switching to cleaner technologies. In addition, the use of both recharging and reuse are having significant impacts in changing unsustainable consumption and production patterns.

Challenges

Priority constraints to implementing effective programmes in this area include:

- Lack of general awareness
- Lack of technical knowledge
- Lack of financial assistance
- Lack of infrastructure

Capacity-building, Education, Training and Awareness-raising

The Government of India has been trying to create an awareness towards moderation of demand and the adoption of a consumption pattern which would not leave a deleterious impact on the environment. The Government, for example, has embarked on an extensive awareness campaign through the print and television media to stress the need to save scarce water, energy, and petroleum resources. This is in conformity with the importance given by Mahatma Gandhi in his thinking on nation and character building. Programmes for policy makers, industries, and/or consumers designed to educate and raise their awareness for more sustainable consumption and production patterns include:

- Training and awareness programmes on Ecomark Schemes
- Training and awareness programmes on Comparative testing
- Training and awareness programmes on BIS
- General Consumer awareness.

Awareness campaign programmes to promote sustainable consumption patterns are carried out through

Programmes of Quality Council of India/BIS/MOEF/CPCB/ Consumer Protection Councils.

Information

The kinds of national information available to assist both decision-makers and industry managers to plan and implement appropriate policies and programmes in this area include:

- Status
- Data
- Notifications
- Standards
- Policy documents

There is a monitoring system is in place to oversee enforcement of relevant laws, regulations and standards. This is carried out by the State Pollution Control Boards (SPCB), Regional offices of SPCBs, and Regional offices of the Ministry of Environment and Forests.

Environment indicators are being developed under the Environment Management Capacity Building Project.

Research and Technologies

Major research and pilot projects and activities are underway in the following areas:

- Life Cycle Assessment (LCA)
- Green Rating

Clean and environmentally sound technologies are promoted and applied through the following means:

- Statutory requirements
- General awareness
- Creating demand for sustainable projects
- Encouraging industries
- Creation of data bank on cleaner techniques
- Creating of infrastructure

Other technology-related issues that are being addressed, in this area are

- Involvement of Fis for financial assistance
- Development of standards
- Capacity building
- Infrastructure awareness

Financing

Activities in this area are financed by the national budget and through external assistance.

Cooperation

Cooperation is carried out between the Government and the World Bank, UNDP, ADB and through bilateral arrangements with a few countries in order to further activities related to promoting sustainable consumption and production patterns.

This information was provided by the Government of India to the 5th and 7th Dessions of the United Nations Commission on Sustainable Development. Last Update: April 1999.

FINANCING

Decision-Making: Coordinating Bodies

In India, the Government has embarked on a macro economic stabilization programme since 1991. Structural reforms in the foreign trade and payments regime, the tax system, industrial policy, and the financial sector have been undertaken, all of which are likely to have implications for the environment. While the Government is attempting to raise resources internally for sustainable development, the importance of international assistance cannot be minimized.

Decision-Making: Legislation and Regulations

In respect of investments including foreign investments, the entrepreneurs are required to obtain Statutory clearances relating to Pollution Control and Environment for setting up an industrial project. A Notification issued under the Environment (Protection) Act, 1986 has listed 29 projects in respect of which environmental clearance needs to be obtained from the Ministry of Environment & Forests. This lists includes industries like petro-chemical complexes, petroleum refineries, cement, thermal power plants, bulk drugs, fertilisers, dyes, paper, etc. However, if investment is less than Rs.500 million, such clearance is not necessary, unless it is for pesticides, bulk drugs and pharmaceuticals, asbestos and asbestos products, integrated paint complexes, mining projects, tourism projects of certain parameters, tarred roads in Himalayan areas, distilleries, dyes, foundries and electroplating industries. Further, any item reserved for the small-scale sector with investment of less than Rs.10 million is also exempt from obtaining environmental clearance form the Central Government under the Notification.

Powers have been delegated to the State Governments for grant of environmental clearance for certain categories of thermal power plants. Setting up industries in certain locations considered ecologically fragile (e.g. Aravalli Range, Coastal areas, Doon Valley, Dahanu etc.) are guided by separate guidelines issued by the Ministry of Environment & Forests.

Decision-Making: Strategies, Policies and Plans

India's Eighth Five Year Plan was drafted in the context of severe resource constraints, and a serious balance of payments situation. The sources of financing projected in the Plan differ from earlier plans in that it seeks to reduce dependence on borrowing, domestic as well as foreign, and on deficit financing, placing greater reliance on resource mobilization and economy in government expenditure. The Eighth Plan contemplates an investment of Rs.7,980 billion (US$ 266,000 million). Considerable reliance is placed on savings of the household sector.

Programmes and Projects

The process to identify environmentally unsustainable subsidy is ongoing and measures are being taken to phase them out. Last year, the Government of Kerala had agreed to phase out subsidies to pulp wood producers over a period of five years.

Status

Implementation of sustainable development programmes as detailed as Agenda 21, requires large amounts of investment. The United Nations Conference on Environment and Development (UNCED) Secretariat estimated that implementation of all activities under Agenda 21 during 1993-2000 would require additional resources of US $ 125 billion a year. This is in addition to the US $ 500 billion a year from National governments and the private sector in developing countries to put their countries on a sustainable development path. The figure was arrived at by estimating the cost of addressing sector and resource specific environment and development problems.

Challenges

At the national level, the other apparent funding mechanism is budgetary support by developing countries for environment protection programmes. However, public expenditure has its limitations. Developing countries, with their limited domestic savings rely on external finances to supplement their resources and overcome budgetary constraints. With the far from favorable trends in external financing, the ability of developing countries to undertake large-scale public expenditure in this field is doubtful. Debt servicing commitments further aggravate the situation. Besides, many developing countries are undertaking economic policy reform, especially fiscal consolidation, and are faced with even more stringent budgetary constraints. At best, only a modest reallocation of resources is feasible.

Information

Information related to financing sustainable development is made available to potential users through: http://finmin.nic.in.

Financing

India has always emphasized the importance of public investment for sustainable development by providing fiscal concessions and incentives. Since energy-efficient technologies and non-conventional energy technologies directly improve the protection level of the atmosphere, several tax concessions, 100% depreciation allowance, and investment subsidies have been made widely available. Investments under the National River Action Plan on Control of River Pollution arising from both municipal and non-municipal waste also produce a major impact on marine and ocean-based resources since they control land-based sources of marine pollution in India. However, additional resources need to be made available through external sources for implementing various programmes and activities listed in Agenda 21.

Cooperation

The goal of Agenda 21 was in part to raise additional external funds for sustainable development activities by increasing bilateral and multilateral Official Development Assistance (ODA) to 0.7% of GNP from donor countries. The fact remains that many of the developing countries are experiencing a net outflow of resources. The average ODA in the post-Rio period 1993-95 has been lower than in the period 1990-92, both in absolute terms and as a percentage of GNP. In fact, ODA at an average of 0.29% of GNP in the 1993-95 period has been the lowest in decades. The Global Environment Facility (GEF) is the only new funding mechanism made available to meet the additional needs identified in Agenda 21. The amount of about US $ 2 billion from GEF, besides the Montreal Protocol Multilateral Fund to tackle ozone depletion, is almost negligible and has fallen short of even the most conservative estimates of the requirements for implementing Agenda 21.

While outlining the estimates of financing needs, Agenda 21 fails to identify the mechanisms to ensure their delivery.

Discussions at the earlier meetings of the Commission on Sustainable Development (CSD), and in the Finance Working Group, have developed a very useful framework for identifying new and innovative sources of funding, including a sectoral approach to mobilizing funds from within the economy and from external sources. Several of the alternatives highlight the important links between the creation of incentives for the reduction of pollution and wasteful consumption in the North and potential financing for sustainable development in the South. More research work on the formulation of such policy options needs to be undertaken to consolidate the progress achieved and to address the unresolved issues.

TECHNOLOGY

Decision-Making: Strategies, Policies and Plans

The facilitating role of Government has been increasing through, for example, the identification of and support for the development of environmentally sound technologies such as chlorofluorcarbon (CFC) alternatives, clean coal technologies, energy efficient technologies, and others. In the field of environmentally sound technologies, a number of research and development projects have been identified for support. The Ninth Five Year Plan projections have stressed the initiation of measures for reducing the energy intensity in different sectors through changes in technology and industrial processes. A critical mass of R&D capacity is crucial for effective dissemination of environmentally sound technologies and their generation locally. Areas which need attention are access to information on state of art technologies, a framework for dissemination of information on the source and availability of environmentally sound technologies, development of guidelines for the transfer of technologies, and training of personnel to undertake technology assessment for the management of such technologies.

Technology upgrading requires that Indian enterprises of all types have information on relevant technologies in international markets and within the country. Many countries have well-developed systems, computerized on-line technology information and dissemination services, often backed by consultancy and financial assistance to enable small and medium enterprises to find out, test, and implement new technologies.

Indian technology policies are undergoing significant changes, and on the whole have improved greatly in recent years. They are not, however, ideal. A coherent technology strategy in India must address a number of interconnected elements in the incentive regime, and the relevant markets and institutions. Technology development generally requires the setting up of clusters of industries that can share information and skills, as in science parks or dedicated industrial estates. Some such facilities exist in India, but their efficacy and functioning need to be strengthened.

Programmes and Projects

In order to strengthen the technological capabilities of Indian industries, both for meeting National needs and for global competitiveness, a number of new initiatives have been launched. A Technology Development Board was established in 1996 with a mandate to facilitate development of new technologies, and the assimilation and adaptation of imported technologies by providing catalytic support to enable industries and R&D institutions to work in partnership with each other. Matching grants to R&D institutions showing commercial earnings through technology services were also introduced in 1996 and will be continued and broadened. Already, a long-term perspective called Technology Vision for India 2020 has been prepared which could form the basis of technology development programmes.

Status

In India, there is considerable technological activity in a wide spectrum of firms. What is most impressive is the number of small and medium sized enterprises that are investing in new technology-based ventures, and often striking out in world market as exporters. However, the rest of the industrial sector still needs to invest in technology upgrading. The experience of many developing and industrialized countries suggests that a rapid acceleration of industrial technology development calls for a deliberate 'strategy', in the sense that it requires the government to coordinate and guide an essentially market-driven process.

Technology development calls for both general and specific forms of human capital, and emerging technologies are highly skill intensive in both technical and managerial terms. While India is endowed at present with large amounts of high-level human capital, investments in the creation of new skills (as measured by enrolment levels in technical subjects at all levels) are low. In addition, firm investments in training are highly variable, with large segments of industry investing very little in training. The small and medium enterprise (SME) sector in particular suffers from very low levels of skill, while industrial training institutes are often unresponsive to their needs. R&D in Indian industry has been rising, but the overall level is still low and over three-quarters of the research effort originates from the public sector. The Government is undertaking an analysis of current technological trends in industry in order to formulate appropriate policies to encourage R&D.

India has a large infrastructure of technology support institutions, some of which are undergoing reform to make them more relevant to industrial needs. A number of universities, especially the Indian Institutes of Technology (IITs), are increasingly interacting with industry on technological matters, while others are outside this circle.

Environmentally sound technologies are essential to achieving sustained economic growth and sustainable development. They encompass a total system which includes know-how, procedures, goods, and services. Agenda 21 emphasizes the need for access to and the transfer of environmentally sound technologies to developing countries on favourable and preferential terms as mutually agreed.

This would take into account the need to protect Intellectual Property Rights as well as the special needs of developing countries for the implementation of Agenda 21. The implementation of the commitments on the transfer of environmentally sound technologies and technical know-how has been disappointing.

Issues of natural resource conservation and agricultural growth cannot be effectively tackled in the absence of an appropriate technological base. In addition, technology is essential for increasing the competitiveness of the Indian economy in international markets. Indigenous development of technology is therefore of the highest importance and deliberate planned steps need to be taken to increase National technology self-sufficiency. Rapid technical progress is altering fundamentally the skills, knowledge, infrastructure, and institutions needed for the efficient production and delivery of goods and services. So broad and far-reaching are current technological developments that many see the emergence of another industrial revolution driven by a new technological "paradigm". This paradigm involves, not only new technologies and skills in the traditional sense, but also different work methods, management techniques, and organizational relations within firms. As new transport and communications technologies shrink international 'economic space', it also implies a significant reordering of comparative advantage, and trade and investment relations between countries.

Challenges

There is a need to strengthen Technology Foresight Programmes to analyze the implications of emerging technologies, domestic strengths and weaknesses, and target future technologies for local development.

Information

In India, the Department of Science and Technology has played an important role in terms of institutional support for building National strengths in scientific fields, and technology assessment and forecasting. A number of technology status reports on energy efficiency, environmentally sound technologies for pollution control, and many other areas have been published.

Programmes and Projects

In India, the Department of Biotechnology has constituted 16 Task Forces for the generation of R&D projects for the development of biotechnologies/techniques/processes, the perfection of techniques/technologies developed, and their field evaluation and transfer to industries for commercialization. This is to benefit the country in general and the affected population in particular. These Task Forces are: aquaculture and marine biotechnology; animal biotechnology; biological control of plant pests, diseases, and weeds; biotech process engineering and industrial biotechnology; basic research in biotechnology; biotechnology based programme for the Scheduled Castes and Scheduled Tribes population, and weaker sections of society; biofertilizer; crop biotechnology and plant molecular biology; environment and conservation biotechnology; food biotechnology; human genetics; medical biotechnology; microbial biotechnology; medicinal and aromatic plants biotechnology; plant tissue culture; and sericulture biotechnology.

These Task Forces consist of experts in the respective areas from different parts of the country. All the Task Forces have identified the needed thrust areas in the Indian context. Based on the recommendations of the Task Forces, the Department has supported research in the following areas: a) development of stress resistant plant species (crop species, forest trees, and medicinally important plant species) for higher yields with less inputs; b) transgenic crop plants for higher yields, pest management, reduction in toxin contents in some crop varieties, etc.; c) development of biological pesticides using biotechnological tools to bring down the pollution load of chemical pesticides; d) development of more efficient bio-fertilizers which will be economical to farmers compared to chemical fertilizers, and ultimately bring down pollution load of chemical fertilizers; e) development of new immunodiagnostic tools for detection of communicable diseases and certain physiological states, such as early detection of pregnancy, etc.; f) development of new/recombinant vaccines for the control of different diseases; g) development of new strains for improved production of antibiotics using/strengthening the existing infrastructure; h) development of highly efficient strains for the treatment of waste waters (domestic as well as industrial) and conversion of wastes and agro-residues into useful chemicals for industrial applications; i) development of ELISA, phase conjugate reflectivity (PCR) techniques, and DNA probes for the detection of enteric pathogens in drinking water so as to avoid epidemic outbreaks by quick corrective measures to be taken immediately after identification of the enteric pathogens; j) development of cleaner technologies using the biotechnological tools; k) development of biosensors for the detection of xenobiotics in the environment; l) conservation of endangered/threatened plant species which are at the verge of extinction and are of economically/ medicinally importance using biotechnological tools, specifically for the development of protocols for ex situ conservation, which would be taken up by the Ministry of Environment and Forests for in situ conservation; m) establishment of gene banks in different parts of the country; n) development of high-yielding technology packages for aquaculture including feed development, breeding and seed production, and bioactive compound, health, development of spawning agents, etc.; and o) development of embryo transfer techniques, animal feed for high mulching cattle and for development of vaccines and diagnostics for different diseases in the area of animal biotechnology.

The purpose of gene banks is for the preparation of an inventory of important species, preservation of genetic resources, and to optimize their uses. There is also a provision for networking of gene banks on a regional or inter-regional basis. Under this programme, banks have been established at: the National Bureau of Plant Genetic Resources (NBPGR), the Indian Agricultural Research Institute (IARI), New Delhi; the Central Institute of Medicinal and Aromatic Plants (CIMAP), Lucknow; and the Tropical Botanical Garden and Research Institute (TBGRI), Thiruvanathapuram.

In collaboration with the Ministry of Environment and Forests, the Department has taken the lead responsibility for the following three activities in the context of the Convention on Biological Diversity: a) development of modalities for access to and transfer of technology to identify the institutions and develop measures for receiving such technologies and utilizing them; b) development of modalities for priority access to biotechnology results and benefits on mutually agreed terms; and c) formulation of procedures for advanced informed agreements on the safe transfer of genetically modified organisms (GMOs) beyond the National jurisdiction.

The recombinant DNA (rDNA) technology heralded new opportunities for beneficial applications in agriculture, animal and human health, industry, and environment. It has also given rise to concern over possible unknown hazards from bridging the natural species barrier and the uncertain effects of new organisms on environmental and public health. In order to have effective and safe release programmes, it is necessary to have biosafety and regulatory arrangements in biotechnology. Realizing the immediate needs for these arrangements, the Department of Biotechnology has prepared the Recombinant DNA Safety Guidelines and Regulations. These guidelines cover the areas of research involving: a) GMOs/living modified organisms (LMOs); b) genetic transformation of plants and animals; c) rDNA technology in vaccines and bioactive molecule development; and d) large scale production and deliberate/accidental release of organisms, plants, animals, and products derived from rDNA technologies.

INDUSTRY

In order to strengthen the technological capabilities of Indian industries, both for meeting National needs and for global competitiveness, a number of new initiatives have been launched. A Technology Development Board was established in 1996 with a mandate to facilitate development of new technologies, and the assimilation and adaptation of imported technologies by providing catalytic support to enable industries and R&D institutions to work in partnership with each other. Matching grants to R&D institutions showing commercial earnings through technology services were also introduced in 1996 and will be continued and broadened. Already, a long-term perspective called Technology Vision for India 2020 has been prepared which could form the basis of technology development programmes.

Programmes and Projects

Adoption of Cleaner Production Technologies and formation of Waste Minimization Circles are being encouraged to minimize environmental pollution. Under the World Bank aided Industrial Pollution Control Project, technical and financial assistance is provided for establishing Common Effluent Treatment Plants (CETPs) in clusters of small scale industrial units at, for example, Mumbai, Surat, and Chennai. An Eco-Mark scheme has been launched to certify various products of industries which fulfil the prescribed standards of environment-friendly production, packaging, and waste disposal.

Status

The expanded scope for specializing in areas of comparative advantage is manifest in the improved growth performance of the economy. Furthermore, while exports have vigorously responded to the removal of the anti-export bias of a protectionist environment, domestic industry appears to have been stimulated by the expanded availability of imports and capital goods, and the challenge of competing in the international market place. The positive response of Indian industry to deregulation is amply demonstrated by the capital goods sector. The capital goods industry which witnessed a negative growth of 12.8% in 1991-92, registered an average growth of about 23% during 1994-96.

Major industrial cities along the coastline like Mumbai, Surat, Cochin, Chennai, Visakhapatnam discharge approximately 0-7 x 109 cubic metres of waste. There are 1,551 industries located along the coastline of the country and all the major ones treat their effluents before disposal. There are innumerable small and medium scale industries which dispose untreated waste into creeks as well as sewage. Several tanneries located in Calcutta and near Chennai have been strongly recommended for closure in cases where they do not install treatment plants within the stipulated period specified by the Apex Court of the country.

This information was provided by the Government of India to the 5th session of the United Nations Commission on Sustainable Development. Last Update: 1 April 1997.

TRANSPORT

Decision-Making: Coordinating Bodies

The various Ministries/agencies involved are those of:

- Surface Transport
- Heavy Industry
- Civil Aviation
- Shipping, Railways
- Urban Development
- Petroleum
- Environment and Forests
- Planning Commission for formulating plans and policy.

The Planning Commission has been entrusted with task of formulating the Integrated Transport policy for the country. As per present system, amendment in existing

Acts/notifications are carried out by the administering Ministry /Department in consultation with the other concerned Department/Ministries.

Planning Commission is finalizing an Integrated Transport Policy which would focus on coordinated and integrated development of all modes of transport.

The Ministry of Surface Transport deals with Indian Motor Vehicle Act, 1988 and Central Motor Vehicle Rules, 1989. Enforcement of these is done by the concerned State authorities. States play an active role in the decisions involving transport and their inputs are considered at all levels of decision making

Decision-Making: Legislation and Regulations

To mitigate pollution problems in Indian cities, more stringent norms for vehicular emissions have been notified under the Central Motor Vehicle Rules which came into effect in April 1996.

Ministry of Surface Transport deals with Indian Motor Vehicle Act, 1988 and Central Motor Vehicle Rules, 1989. The Motor Vehicle Act (Amendment), 2000 legislated the use of environment-friendly fuel like Compressed Natural Gas (CNG) and Liquefied Petroleum Gas (LPG) as auto-fuels. In addition, Bharat Stage-I norms, which are akin to Euro-I norms, have been introduced all over the country w.e.f. 1.4.2000. Further, Bharat Stage-II norms, which envisage a Sulphur content of 0.05% as against higher quantities under Bharat Stage-I, have been introduced in NCR of Delhi, and these are being introduced in the other three metropolitan cities namely Mumbai, Calcutta and Chennai in a phased manner. For Railways the Railways Act 1989 is administered by the ministry of Railways.

Bharat Stage-I norms have been introduced all over the country w.e.f. 1.4.2000. This norm is being progressively tightened and Bharat Stage-II norms which envisage a Sulphur content of 0.05% maximum for petrol and diesel both, have been introduced in Delhi from 1.4.2000 and in Mumbai from 1.1.2001 and in both Chennai and Calcutta from 1.7.2001. The ministry of Finance provides tax incentives for popularizing use of environment-friendly fuel like CNG, LPG etc. Chairman, Central Pollution Control Board is chairing an inter-Ministerial Committee to specify the road map for future emission norms for the country. The report is expected by March 2001.

Ministry of Railways has its own codal procedures, safety norms, rules and regulations for all aspects of Railway working including Safety, Operations and Maintenance of assets.

In August 2000, use of environmental-friendly fuel like CNG, LPG, fuel cell etc. for automotive purposes has been legislated. Relevant rules are under formulation in respect of LPG. In case of CNG, there are existing guidelines. In addition, Ministry/o Petroleum as well as Department of Explosives are also amending their respective regulation in this regard.

Decision-Making: Strategies, Policies and Plans

A Task Force has been set up in the Planning Commission for the development of an integrated transport policy. The policy would focus on integrated development of transport sector, interconnectivity of various modes of transport, financing plans of all modes of transport and other issues.

To address the diverse issues facing the transport sector the need for a comprehensive policy package has been recognized Government has drawn a plan to strengthen the Indian railway system in its reach and capacity so that it effectively links the distant parts of the country, helps to develop the economic potential of the remote areas and carries the bulk of the nation's long or medium haul traffic. Similarly, the road network is being expanded and strengthened to improve accessibility of the hinterland, especially the rural areas and to facilitate the integration of the isolated parts of the country. The length and breadth and the quality of the highways has improved greatly as part of a national grid to provide for speedy, efficient and economical carriage of goods and people.

Government is making efforts to regulate road transport for better energy efficiency and pollution control, and to make the mass transport network viable through a rational tariff policy and a refurbishment of the fleet. The capacity of the ports in terms of their berths and cargo handling equipment are being improved to cater to the growing requirements of the overseas trade. The shipping industry needs to be enabled to carry higher shares of the sea-borne trade in indigenous bottoms. The civil aviation sector is being expanded to increase its carrying capacity for passengers and cargo, improve the ground handling facilities and provide connectivity to areas like the North-East. Conditions need to be created to ensure full utilization of the capacities created in the public sector with large investments made in the past.

In the metropolitan areas, on the one hand, the provision of mass public transport is being increased through a mix of environment-friendly modes - specially designed buses, light rail and metro and on the other, demand management is being ensured through price-based as well as non-price-based measures so as to minimize the dependence on personalized transport. Similarly, non-mechanized transport should be accorded its rightful niche in a well-conceived transport network.

To bring about this sea change in the transport scene, many policy initiatives will be needed, each backed by adequate investment and complemented by suitable policy changes in other sectors. A Task Force on Infrastructure has been constituted with the aim of attracting investments to specific projects.

TASK FORCE ON INFRASTRUCTURE

- A Task Force on Infrastructure was constituted under the Chairmanship of Dy. Chairman, Planning

Commission comprising both Government and industry representatives with the aim of attracting investment to specific projects of national and regional importance, and ensuring their timely completion.

Initially, the Task Force will deal with the following projects focusing on innovative methods for financing them.

- Six lane expressway of 7,000 km length, having North-South and East-West corridors.
- Four-laning of National Highways, and
- Five international airports.

The terms of reference of the Task Force include:

- determining the routes for Expressways and National Highways, and establishing technical parameters thereof;
- identifying and recommending locations for the airports;
- establishing benchmarks and criteria for the airports;
- recommending financing options for Expressways, highways and airports;
- recommending criteria for competitive bidding and selection of (EPC) contractors;
- recommending measures as are necessary for timely completion of projects including governmental clearances; and
- overseeing and monitoring timely implementation of the projects.

The Task Force will also formulate an Integrated National Transport Policy to strengthen the transport infrastructure in the country. It would also recommend steps that can significantly improve and foster reforms in those key segments of the economy.

The following are the Strategic Objectives of the Civil Aviation policy of India

The objectives of this policy are the creation and continued facilitation of a competitive and service-oriented civil aviation environment in which:

i. the interests of the users of civil aviation are the guiding force behind all decisions, systems and arrangements,

ii. safe, efficient, reliable and widespread quality air transport services are provided at reasonable prices,

iii. there exists a well-defined regulatory framework catering to changing needs and circumstances,

iv. all players and stakeholders are assured of a level playing field;

v. private participation is encouraged and opportunities created for investors to realize adequate returns on their investments;

vi. recognizing that aviation today is an important element of infrastructure, rapid upgradation of airport infrastructure with priority to the busiest airports and those handling international flights;

vii. recognizing that transportation of air cargo is vital to the economic growth of the country, creation and development of specific infrastructure for air transportation of cargo and express cargo is encouraged,

viii. "airline operations and acquisition of aircraft" is conferred "infrastructure" status for overall growth of the civil aviation sector in the country

ix. domestic and international aviation in the country are encouraged to grow at par with world aviation industry;

x. inter-linkages with other modes of transport are encouraged and stimulated;

xi. trade, tourism and overall economic activity and growth is encouraged;

xii. international cooperation in aviation and development in tune with international trends and best practices, consistent with airspace sovereignty is promoted;

xiii. indigenous development of aircraft, components and aviation products is encouraged,

xiv Security of civil aviation operations is ensured through appropriate systems, policies, and practices, and xv Effective systems are put in place for timely crisis and disaster management, including investigation of incidents/accidents.

Decision-Making: Major Groups Involvement

India is a democratic country and the elected representatives of the people are invariably connected with the decision-making process. Their participation at the National level is through Parliamentary proceedings, meeting of the various Committees of Parliament as well as individual initiatives. States and local Governments have corresponding arrangements. The inputs of various organized Bodies/ Associations are also considered during policy formulation. Parliamentary Committees such as the Standing Committee, Railway Convention Committee hold regular meetings with the Railway Administration regarding overall functioning of the Railways. Representations, suggestions and recommendations from various quarters such as the Passenger Amenities Committee, Zonal Rail Users Committee and Trade Associations like the CII, ASSOCHAM and FICCI are also considered.

In the urban areas, lack of adequate mass transport, complete absence of demand management and policy distortions in the area of fuel pricing and bank finance have resulted in an explosion of personalized transport comprising mainly scooters and cars. This has contributed to high levels of pollution and alarming rates of accidents. On the other side, a large number of villages lack a reliable all-weather connection with nearby markets and towns. Certain environment-friendly and socially cost-effective means of transport like coastal shipping, inland water transport and non-mechanized transport, human or animal-powered have remained undeveloped.

Planning for, and execution of Railway Projects is done in an integrated and need-based manner keeping a national perspective in view. Geographical boundaries do not form a criterion for determining projects.

The Union Government and number of State Governments have approved the participation of private sector for construction, maintenance and operation of Highways on a BOT basis. A number of projects are under implementation. Government is encouraging and framing laws and regulations for promotion of private sector participation in various areas of transport including roads and civil aviation.

Programmes and Projects

Efforts are being made to make the roads more friendly for road users through a series of measures like widening of roads, construction of by-passes, improvements of road geometric providing paved shoulders, strengthening weak pavements, replacement of level crossings by road over bridges, retro reflective road signs, provision of wayside amenities on high traffic density corridors and creation of awareness amongst various categories of road users.

Traffic efficiency of the Railways is being achieved through the following programmes:

- Use of more efficient locomotives;
- Improving wagon productivity;
- Increasing the axle-load;
- Upgrading Signal and Telecom; and
- Modernization of Terminals.

Planning Commission has constituted a high level working group to look into various aspects of road safety. In this respect, the following research project is sanctioned by the Ministry of Road Transport and Highways is in progress and the draft final report is under examination. Besides the Indian Road Research Institute is working on some projects.

RSDO (Research Designs & Standards Organization) is under the Ministry of Railways, which takes care of the research needs of the Indian Railways through Mission Programmes. Some of the missions of RDSO include Heavy Haul Operation and High Speed Trains.

To reduce the use of fossile fuel, the Government has launched a programme for alternative modes of surface transportation. The programme has demonstrated the use of electrical vehicles in major metro cities with encouraging signs of interest from a number of private companies who would like to manufacture such vehicles. A new programme to replace petroleum by methanol has also been launched by the government.

Status

The domestic funding of the transport sector can be either public or private. Historically, the investments in the transport sector, particularly in the rail, road, ports and airports infrastructure, have been made by the State mainly because of the large volume of resources required, long gestation periods, uncertain returns and various externalities, both positive and negative, associated with this infrastructure. However, the galloping resource requirements and the concern for managerial efficiency and consumer responsiveness have led to the active involvement of the private sector in infrastructure services in recent times.

In India too, considerable private investment exists in trucking, inter-city bus travel, shipping and lately in airline services. By and large, mobile transport units like trucks, buses, wagons, ships and aircrafts lend themselves easily to private investment while the large fixed infrastructure has remained in the domain of public investment. However, statutory and administrative initiatives have been taken in recent years to involve private capital in the expansion and strengthening of infrastructure in the railways, road, shipping and the airports. Private participation can take many forms like full or joint ownership management contract, leasing, and concessions like BOT. However, looking at the state of Indian capital markets, particularly for long-term debt, it may not be realistic to expect any large-scale contribution from the private sector in the transport area though evolution of a well-developed policy and regulatory framework can result in gradual inflows from the private sector.

The scope for private sector participation in providing rail infrastructure and services is limited. Attempts have, however, been made to involve the private sector in augmenting the capacity of the railway system in a number of ways. The "Own Your Wagon Scheme (OYWS)" was launched in the Eighth Plan in order to tap the private sector resources for augmenting the supply of wagons. The private sector firms would procure the wagons, own them and lease them to the railways with or without preferential claim on allotment of capacity for the firms' own use. With the revision made in the scheme in February 1994, the response has been encouraging during the period 1994-95 to 1996-97. The actual procurement under the scheme has been of the order of 9000 i.e. about 77% of the target.

It has also been decided to undertake some of the projects through investment by the private sector under the "Build-Own-Lease-Transfer (BOLT)" scheme -the private entrepreneurs and the financial institutions would whereby build/manufacture/finance the assets for lease to the railways. It is proposed to offer projects like gauge conversion, supply of rolling stock, electrification, doubling of existing single lines, and telecom projects under this scheme. However, the experience during the Eighth Plan has underlined the difficulty of enlisting private participation in such infrastructural areas on affordable terms. In order to attract larger private sector investment in the Railways, maximize the efficiency and benefit from the synergistic effect of the private and public sector investment, restructuring of Indian Railways becomes inevitable.

Various fiscal and tax concessions have been offered for undertaking road projects on a BOT basis. The Government has also assured help in land acquisition, environmental clearances and simplification of procedures. Some State Governments have also taken significant steps like the setting up of dedicated organisations on the pattern of NHAI to promote road development, financed through user charges on BOT or State toll basis. Some small stretches of roads and bypasses have already been commissioned under these arrangements. The role of this device may become more significant once the policy framework including an independent regulatory system is put in operation and investors are able to overcome their apprehensions with regard to uncertainties of these long-term investments in such infrastructure. The policy framework is sought to be kept flexible to permit allocation of risks on an acceptable basis and sharing of risks between the Government and the private sector through grants or equity sharing.

Projects relating to bypasses, bridges and 4 laning of existing sections of National Highways are financially viable and bankable would be taken up through private sector participation. 11 projects involving an investment of about $120 mn have already been initiated under the BOT Schemes. As stated already, viable sections o the four major corridors are also proposed to be taken up under the BOT scheme for four laning. NHAI has been given considerable flexibility to financially collaborate with the private or the public sectors and projects which are not viable on the basis of traffic density will be provided equity/loan support from NHAI. Several other ROB/Bypass projects have been proposed under the BOT scheme through State Governments. In addition to the projects taken up through private sector participation under the programme of NHAI, some investment would be available under privatization programme for non-NHAI roads. Thus, given the risk profile of toll road projects and the relative under development of Indian long-term debt market, private investment in roads is expected to be modest. Therefore, the budgetary support will continue to have a large and crucial role to play in road development.

An overview of availability / accessibility is as following:

RAILWAYS

The railways in India provide the principal mode of transportation for freight and passengers. It brings together people from the farthest corners of the country and makes possible the conduct of business, sight-seeing, pilgrimage and education. Indian Railways have been a great integrating force during the last hundred years. It has bound the economic life of the country and helped in accelerating the development of industry and agriculture. From a very modest beginning in 1853, when the first train steamed off from Bombay to Thane, a distance of 34 km, Indian Railways have grown into a vast network of 7,068 stations spread over a route length of 62,495 km with a fleet of 7,206 locomotives, 34,728 passenger service vehicles, 5,302 other coaching vehicles and 2,63,981 wagons as on 31 March 1998. The growth of Indian Railways in the 145 years of its existence is thus phenomenal. It has played a vital role in the economic, industrial and social development of the country. The network runs multi-gauge operations extending over 62,495 route kilometre. The gauge-wise route and track lengths of the system as on 31 March 1998 were as under:

Gauge Route Running Total km track

Broad Gauge (1,676 mm) 43,083 60,550 83,073

Metre Gauge (1,000 mm) 15,804 16,682 20,343

Narrow Gauge (762 mm and 610 mm) 3,608 3,676 4,027

Total 62,495 80,908 1,07,445

About 22 per cent of the route kilometre, 30 per cent of running track kilometre and 29 per cent of total track kilometre is electrified. The network is divided into nine zones and further sub-divided into divisions.

ROADS

India has one of the largest road networks in the world. The country's total road length as on 31 March 1996 stood at 33,19,644 km. The Ninth Plan laid emphasis on a coordinated and balanced development of road network in the country under (i) primary road system covering state highways; (ii) secondary and feeder road system covering state highways and major district roads; and (iii) rural roads including village roads and other district roads.

Substantial outlays were proposed for road development in the rural and tribal areas. During the Eighth Plan, $400 million was spent on Central sector roads and $3000 million spent on state sector roads. Under the Ninth Plan, the allocation is $2000 million for the Central sector roads programme.

NATIONAL HIGHWAYS

The Central government is responsible for the national highway system totaling a length of 51,966 km. In 1947 approximately 2,500 km of missing road links and thousands of culverts and bridges, which did not exist, were required to be constructed to have an integrated and continuous network.

There was an increase in missing road links with addition of new roads to the national highway system in later years. Up to 31 March 1999, road links including diversions realignments, and constructed totaled 4,717 km, widening and strengthening single lane section to double lane carriage was done in 25,685 km, widening to four-lane completed in 845 km, strengthening of weak two-lane pavement has been done in 14,790 km. In addition 560 major bridges and 3,078 minor bridges were also completed. During Eighth Five Year Plan a sum of about $ 800 million has been spent on development of national highways. $350 million and $340 million have been spent on National Highways during 1997-98 and 1998-99 respectively and budget allocation of $510 million has been made for the year 1999-2000. Though the national highways constitute only two per cent of the total road length, they carry nearly 40 per cent of road traffic.

There are altogether nine on-going external loans for the improvement of National Highways, comprising one loan (US dollar 306 million) from the World Bank, three loans (total US dollar 672 million) from the Asian Development Bank (ADB) and five loans (total Japanese Yen 36,915 million approximately equivalent US dollar 450 million) from Overseas Economic Cooperation Fund (OECF), Japan. The World Bank loan includes six National Highways sub-projects in the States of Punjab, Haryana, Orissa, Maharashtra, Madhya Pradesh and West Bengal and one state highway project for reconstruction of bridges in Orissa. All the sub-projects are in progress. The first ADB loan (US dollar 177 million) provides for development of national highways in three States, Andhra Pradesh, Haryana and Uttar Pradesh and state highways in Andhra Pradesh, Karnataka and Tamil Nadu. All works have been substantially completed by March 1998. The second ADB loan (US dollar 250 million) provides for improvement in Karnataka, Kerala, Rajasthan and of state roads in Andhra Pradesh, Orissa, Uttar Pradesh and West Bengal. The NH projects in Andhra Pradesh and Orissa have been substantially completed. Remaining projects are in progress. The third ADB loan covers national highways projects in Andhra Pradesh, Haryana, Rajasthan, Bihar and West Bengal. All the projects are in progress. The project for improvement of Mathura-Agra Section of NH-2 in Uttar Pradesh, under one of the five loans of OECF is in progress. Out of the remaining four loans, the procurement process has been recently completed for two, and is in progress for the other two.

STATE SECTOR ROADS

State highways and district rural roads are the responsibility of State governments and are maintained by various agencies in States and union territories. Roads are being developed in rural areas under the Minimum Needs Programme (MNP). The objective of it being to link all villages with a population of 1,500 with all-weather roads. The Government also assists in development of certain selected roads in States. The total length of roads in India stood at 33,19,644 km (both surfaced and non-surfaced roads) in March 1996.

INLAND WATER TRANSPORT

India has got about 14,500 km of navigable waterways which comprises rivers, canals, backwaters and creeks. At present, however, a length of 3,700 km of major rivers is navigable by mechanized crafts but the length actually utilized is only about 2,000 km. As regards canals, out of 4,300 km of navigation canals, only 900 km is suitable for navigation by mechanized crafts. About 1.60 million tones of cargo is being moved by Inland Water Transport (IWT), a fuel efficient and environment friendly mode. IWT is also known for higher employment generation potential. Its operations are currently restricted to a few stretches in the Ganga-Bhagirathi, Hooghly rivers, the Brahmaputra, the Barak river, the rivers in Goa, the backwaters in Kerala and in the deltaic regions of the Godavari-Krishna rivers. Besides the organized operations by mechanized vessels, country boats of various capacities also operate in various rivers and canals.

CIVIL AVIATION

The civil aviation sector has three main functional divisions—regulatory, infrastructural and operational. On the operational side, Indian Airlines, Alliance Air (Subsidiary of Indian Airlines), private scheduled airlines and air taxis provide domestic air services and Air India provides international air services. Pawan Hans renamed Pawan Hans Helicopters Limited provides helicopters services to ONGC in its offshore operations and to inaccessible areas or difficult terrains. Indian Airlines operations also extend to the neighboring countries, South East Asia and Middle East. India has been a member of the International Civil Aviation Organization (ICAO) and is also on the Council of ICAO since its inception.

The Government has ended the monopoly of Indian Airlines and Air India on the scheduled operations by repealing the Air Corporation Act, 1953. There are at present three private scheduled airlines operating on the domestic network rendering the passengers a wider choice of flights. Apart from this 37 Air Taxi Operators are providing non-scheduled air services. A new policy on domestic air transport service was approved in April 1997 according to which barriers to entry and exit from this sector have been removed;

choice of aircraft type and size has been left to the operator; entry of serious entrepreneurs only has been ensured; and equity from foreign airlines, directly or indirectly, in this sector has been prohibited. The existing policy on air taxi services providing for a route dispersal plan to ensure operation of a minimum number of services in the North-Eastern Region, Andaman and Nicobar Islands, Lakshadweep and Jammu and Kashmir has been retained.

The Airports Authority of India (AAI) manages 92 civil airports including five international airports, 28 civil enclaves at defense airfields. It manages the entire Indian air space

RAILWAY ENERGY CONSUMPTION (1998-99)

All gauges

Electric	602 3918 K.Whs.
Diesel	1900 804 Kilo litres

To reduce emissions from vehicles and to increase the engine efficiency, a number of steps have been taken in India. Apart from introduction of unleaded gasoline, the sulphur content of Gasoline has been reduced from 0.2% to 0.1% wt. in the entire country with effect from 1.4.2000. 0.05% wt sulphur Gasoline is being supplied in NCR, Delhi w.e.f. 1.4.2000 and other metros with effect from 1.10.2000.

Benzene content limit in Gasoline has been introduced from 1.4.2000 as 3% maximum in metros and 5% max in rest of the country. The same is further reduced to 1% max in Mumbai &NCT, Delhi with effect from 1.10.2000 and 1.11.2000 respectively. Supply of 1% Benzene gasoline is planned in NCR, Delhi with effect from 31.3.2001.

To reduce the emission of Sulphur Dioxide from diesel vehicle, sulphur content of high speed diesel (HSD) has been reduced from 1%wt to less than 0.5%wt from 1.4.1996 in four metros and Taj trapezium zone. Sulphur content in HSD was further reduced in the Taj trapezeum and Delhi to 0.25% wt with effect from 1.9.1996 and 15.8.1997 respectively and the same is implemented in all four metros i.e. Delhi, Kolkatta, Mumbai and Chennai from 1.4.1998. With effect from 1.1.2000 0.25% wt SHSD is being supplied in entire country. Further 'S' content is reduced to 0.05% wt in Delhi & Mumbai w.e.f 1.4.2000 & 1.10.2000 respectively and Kolkatta &Chennai w.e.f. 1.1.2001.

In addition to the sulphur content, other important parameters of HSD like Cetane number and distillation specifications have also been improved wit effect from 1.4.2000. Apart from the above, there is proposal to reduce sulphur content of both Gasoline and HSD to 50 ppm in future.

In order to encourage use of eco friendly alternate fuel, a programme has been launched for use of Ethanol Blended Gasoline. For this purpose two pilot plant projects, one each at Maharashtra and U.P have been taken up. Lead-free petrol has been inducted all over the country.

About 65 to 70% of the air pollution in metropolitan cities (Delhi, Mumbai, Chennai and Kolkatta) are attributed to vehicular emissions. The vehicular exhaust emissions include particulate matter, carbon monoxide, oxides of nitrogen, and sulphur dioxide.

In accordance with the 'White Paper on Pollution in Delhi with an Action Plan' two wheelers account for about two-thirds of the total vehicular population in Delhi. Because of inherent drawbacks in the design of two-stroke engines, two-wheelers emit about 20-40 percent of the fuel un-burnt/ partially burnt. Mass emission standards known as India 2000 norms akin to Euro-I norms are effective for all categories of vehicles manufactured with effect from 1.4.2000 in the entire country. Further, Bharat Stage-II emission standards akin to Euro II norms are effective from 1.4.2000 for four wheeled private (non-commercial) vehicles in Delhi and the same is effective in Mumbai from 1.1.2001 and in Chennai and Kolkatta from 1.7.2001 in accordance with the requisite notifications issued by the Ministry Of Surface Transport.

According to the ambient air quality data, there is an improvement in the air quality during the current year as compared to previous years in Delhi.

Challenges

India's is an integrated approach that seeks to improve all the modes -land, sea and air. The surface transport sector requires most attention.

Financial resources, terrain, climate and population pressure are some of the constraints. These issues are beginning to get addressed with the economy registering healthy growth.

Emissions can be reduced through supply of fuel of appropriate specifications, use of appropriate technology in vehicular engines and better inspection and maintenance of on-road vehicles. As regards technology, it is available in the country and can be introduced in a phased manner corresponding to the availability of appropriate quality of fuel. As regards inspection/maintenance, in case of commercial vehicles annual fitness tests are mandatory after the first two years. For non-commercial vehicles, the period is 15 years. The norms of fitness are being progressively tightened. However, due to a stay order granted by the Calcutta High Court, the involvement of private sector in inspection of vehicles is not allowed. This Ministry of Environment and Forests and the Central Pollution Control Board has gone in appeal to the Supreme Court for a vacation of the stay. Involvement of the private sector for inspection and maintenance is essential for improvement in the ambient air quality through reduction in emissions from on-road vehicles.

Transport and traffic systems that are not managed well put people at risk on the health front. Accidents, holdups and polluting vehicles cause problems that effect society.

Capacity-building, Education, Training and Awareness-raising

There are several efforts being made by the Government at all levels to increase public awareness. Public awareness on environmental issues is being taken up on mass scale and have been included in the school curricula. The issues which are included in the school curricula include topics like steps to increase forest cover, control soil erosion and reduce GHG emissions etc. Separately, major campaigns are launched by various schools on environmental issues whereby children try and educate the public on the need to improve the environment. The State controlled television and radio media very frequently feature programmes giving ways and means on how to contain environmental degradation.

A number of seminars, workshops and training programmes are organized by the Government of India for creating awareness about renewable energy among different sections of the society including for policy makers, industries and also for users. These programmes are carried out by State Governments, Academic and R&D institutions, NGOs and industries.

There are no rules at national level regarding non-motorized transport. However, local bodies may prescribe certain rules at their level. Government has incentives like seasonal tickets in public transport system.. Car pools have been formed for senior Government officers. The Government has planned efficient mass rapid transport systems for all metros. In Delhi construction is already underway.

Awareness about road safety is being generated through use of audio-visual media, news media, conducting programmes in schools, campaigns through NGOs as well as distribution of posters, pamphlets, games etc. In addition, refresher driving training is being provided to drivers of heavy commercial vehicles through various NGOs/institutes.

Nature study field visits coupled with audio visual presentations are part of school curricula. Children are made to do projects that are designed to learn more about the environment and its relationship with other sectors. There are many issues on environmental protection in the school curricula.

Various programmes of trainings in the country and abroad for training of Highway/Traffic Experts, technical staff are in operation.

Information

The Ministry of Surface Transport is maintaining the traffic census database for National Highways. The traffic census is made manually by the respective State Public Work Departments and hard copy of data are forwarded to this Ministry. This data is then entered into computers and analyzed.

Railways maintain comprehensive data relating to all aspects of their working. These data are essential tools at the hands of the management in traffic planning, forecasting and fixing targets.

Central Pollution Control Board, a statutory body under the Ministry of Environment and Forests has established 290 ambient air quality monitoring stations covering major cities and urban centres in the country and the data obtained is so processed and evaluated as to recommend necessary mitigative and control measures.

Based on measurements of total suspended particulate matter the air quality in 70 cities during 1997 by Central Pollution Control Board the following categories could be made.

- 29 cities are critically polluted (above 1.5 times the standard*)
- 22 cities are highly polluted (between 1 and 1.5 times the standard)
- 17 cities are moderately polluted (between 0.5 and 1 time the standard); and
- 2 cities with clean air (0.5 times the standard)

National Ambient Air Quality Standards for ambient air quality have been notified under the Air (Prevention and Control of Pollution) Act, 1981 and in the Environment (Protection) Act, 1986.

Research and Technologies

Research, development and demonstration projects in the field of Electric Vehicles (EV's) under Alternative Fuel for surface transportation programme are also underway. The objective of this programme is to develop non polluting EV's with rechargeable batteries and fuel cells as a power source. CNG buses and other forms of less polluting modes of transport are being given incentives.

The strategy of the government tot develop an integrated transport system aims to improve the efficiency of the system.

Railways have a well-devised system of working with regard to Traffic management and increasing safety. The improvement in traffic indices is being achieved through:

- Improved terminal handling.
- Upgradation of Signal and Telecommunication.
- Introduction of new concepts like Engine-on-Train concept.
- Reduced detention to Rolling stock for improved turn-around.

Financing

The main sources of funding for building infrastructure like National Highways are the National budget, cesses collected from sale of petrol and diesel and assistance from external, multilateral and bilateral agencies. In addition, small amount for funding is from private sector.

Cooperation

India is a Member country of ESCAP and is a participating country in the Asian Highways System.

Bilateral agreements for the transportation of goods between India & Bangladesh and India & Pakistan exist. Indian Railways is a member of both UN-ESCAP and UIC

(International Union of Railways). UIC is a voluntary association of World Railways with over 140 members around the world. UN-ESCAP and UIC are currently engaged in the process of developing Trans-Asian Railway Corridor connecting Europe and Far East countries. The proposed corridor passes through Turkey, Iran, Pakistan, India, Bangladesh, Myanmar, Thailand and Singapore. Indian Railways cooperate with both UIC & UN-ESCAP with regard to the Corridor development plan.

India is associated with various international organization in the field of Highways and Management like PIARC, IABSE, IRF. India is also associated with regional groups like SAARC, BIMSTEC.

SUSTAINABLE TOURISM

Decision-Making: Coordinating Bodies

The Ministry of Tourism, Government of India is the nodal agency for the formulation of national policies and programmes and for the coordination of activities of various central and state government agencies and the private sector for development of Tourism. However, all the environmental regulations are enforced by the Ministry of Environment and Forests.

State Governments/District Administration/Local Bodies and Councils are responsible for sustainable tourism at the local level. The Forest Departments of the respective State Governments and the Union Ministry of Environment and Forests enforces environmental regulations.

Decision-Making: Legislation and Regulations

There are several Acts and laws which ensure sustainable tourism. These are the Wild Life Protection Act 1972, the Environment (Protection) Act 1986, and Prevention of Cruelty to Animals Act 1986. These do not set aside any specific area for tourism, but such areas have to be identified by the State Governments and obtain the required approvals/ relaxations.

There are established procedure stipulated by the Ministry of Environment and Forests for project clearance and monitoring. There are deterrents in these strategies to check, control or penalise damaging environmental practices on the part of businesses and visitors

Decision-Making: Strategies, Policies and Plans

The National Policy on Tourism lays emphasis on sustainable development of tourism. In addition, the Government has brought out a comprehensive Eco Tourism Policy and Guidelines. All issues relevant to sustainable tourism are covered in the policy and guidelines. The Eco tourism policy and guidelines will ensure regulated growth of eco tourism and nature based tourism with its positive impacts of environment protection and community development.

Ecotourism policies and Guidelines have been formulated by the Government in consultation with the industry and are being implemented on a voluntary basis. The environment regulations are mandatory. All the players of the tourism industry including consumers have hailed these codes and have shown sensitiveness to the environment. The National Policy on tourism lay emphasis on development of ecotourism and preservation of natural and man-made tourist attractions and resources. The policies and guidelines on ecotourism are specifically on the development of sustainable tourism.

Decision-Making: Major Groups Involvement

The National Tourism Policy envisages a very big role for all the stakeholders in the decision making process. They are involved in the development of tourism and have contributed substantially in sustainable tourism in the country.

Programmes and Projects

.The major programmes to promote sustainable tourism include:

- Preparation of Master Plans for tourism development
- Tourism Awareness
- Integrated Development of Destinations.
- Implementation of Eco tourism policies and programmes

Examples of the ways in which eco-tourism and nature-based tourism are being promoted include:

- Eco-tourism projects in Kerala-Coconut Grove and Spice Village Resorts.
- The Bangaram Island Resorts
- Bekal Resorts, Kerala
- Jungle Lodges and Resorts, Karnataka
- The sustainable development project of Andamans.

Status

Tourism is second largest net earning of foreign exchange generating about 3 billion dollars. Tourism generates larger employment opportunities per tourist. The direct employment in the sector during 1997-98 is estimated to be about 9.8 million and total employment due to tourism is 23.1 million. The foreign tourism arrivals during 1998 were 1.59 million. It has grown up to 2.37 million during 1997. The tourism arrivals growth is expected to be about 8% for the period up to 2000 AD. The present level of tourist traffic has not impacted any serious problem on sustainable development. However, there is a large movement of domestic tourism within the country Hotels and other tourist establishments follow the guidelines framed by the Ministry of Tourism.

The main strength of Indian tourism at present is its cultural attractions, particularly monuments and archeological remains, fairs and festivals, wild life and beaches. The aim of the tourism policy now is to diversify the tourism product in such a way that the development of ecotourism and nature based tourism is promoted to attract environmentally conscious tourists.

Challenges

The Ministry of Tourism is making vigorous effort for the sustained growth of tourism through synergy programme and establishing effective co-ordination with state governments and other agencies to develop infrastructure for sustainable tourism. There are no serious constraints on pursuing sustainable tourism.

Capacity-building, Education, Training and Awareness-raising

The Indian Institute of Tourism and Travel Management organise periodical training programmes on various aspects of Sustainable tourism. They also organise separate programmes for policy makers and administration. The Ministry of Tourism has launched Tourism Awareness Programmes in collaboration with Pacific Asia Tourism Association (PATA) India Chapter and the Industry. The programme has been well received by the industry and the consumers.

Information

Information is disseminated through brochures, computer media, Workshops for tour operators, travel agents and other concerned parties.

Mapping and inventorying of natural resources and ecosystem characteristics in tourist areas is an ongoing activity of the Government.

Potential users of information have to approach the concerned agencies. Information is also available on the World Wide Web.

Research and Technologies

Technology-related issues are already incorporated in the ecotourism and Environment Policies of the Government.

Financing

Activities in this area are through the national budget and private sector partnership.

Cooperation

The Ministry of Tourism, Government of India is promoting India as a major tourist destination through its tourist offices in abroad and mission. The Ministry is also participating in international meetings/exhibition, fairs and festivals, seminars, etc. Recently, Ministry of Tourism has launched a multimedia CD-ROM on Indian Tourism. Sustainable tourism is promoted through all these efforts. The Environmental pledge is printed in every brochure produced by the Ministry of Tourism.

Bangaram Island in Lakshadweep group of Islands, India is a model sustainable tourism destination. Cooperation is generally provided through guidance and financial assistance. Examples of other cooperation in this area include: The conservation project of Ajanta and Ellora, which has been taken up with the assistance of Overseas Economic Cooperation Fund (OECF) of Japan.

A study for the development of a strategy for Environmentally sustainable Tourism in Andamans has been completed with the assistance of World Tourism Organisation (WHO) and United Nations Development Programme (UNDP).

NATURAL RESOURCE ASPECTS OF SUSTAINABLE DEVELOPMENT IN INDIA AGRICULTURE

Decision-Making: Coordinating Bodies

Land, being a subject under the exclusive jurisdiction of States, there is no national legislation, which restricts transfer of productive arable land to other uses. However, State Governments have enacted legislation on the subject, which provides restriction on use of land for non-agricultural purposes.

Decision-Making: Legislation and Regulations

Statutory environmental clearance under Environment (Protection) Act is required for the following types of agricultural development projects and human settlement projects:

- *Agricultural Development Projects*: Major irrigation projects with command area of 10,000 hectares and more.
- *Human Settlement Projects*: Located in the Coastal Regulation Zone. Statutory environmental clearance is required for 30 selected activities in sensitive areas such as Coastal Regulation Zone, Doon Valley, Dahanu Taluka, Murad Janjira, Numaligarh, Aravalli (identified areas in Gurgaon Taluka of Haryana and Alwar Taluka of Rajasthan).

While examining the proposals, the impact of the projects on different Ecosystems, including agricultural lands are examined. The feasibility of avoiding agricultural land for other developmental activities is also examined. However, there is no statutory restriction in transferring agricultural land for other uses.

In case the lands involved forestland, the project proponent has to obtain clearance under Forest (Conservation) Act for the use of forests for non-forest purposes. In case it involves National parks/Sanctuaries, if the activity is not beneficial to the wildlife, it cannot be taken up in those areas. In regard to human settlements, namely, buildings in the coastal regulation zone, there are restrictions on height, plinth area, drawl of groundwater disturbing the landform, disposal of waste, etc. Further, construction of buildings is prohibited in the sensitive areas within Coastal Regulation Zone.

To ensure availability of effective pesticides, a comprehensive Central Legislation ? Insecticides Act, 1968 - is being implemented. Central Insecticides Laboratory, Registration Committee, Central Insecticides Board and Regional Pesticides Testing Laboratories are the principal wings for implementation of the Act at the Central level. To save the Indian agriculture from exotic pests and diseases, legislative measures on Plant Quarantine are being enforced through 26 Plant Quarantine Stations located at International Airports, Seaports, Land Frontiers. These Stations also discharge the responsibility of phytosanitary certification to help export of agricultural commodities.

In order to promote the use of safer pesticides and also increasing the export potential of pesticides, the Central Insecticides and Registration Committee set up under the Insecticides Act, further simplified data requirements for both plant origin and provisionally registered neem-based pesticides and bio-pesticides.

The main objectives of the Government's price policy for agricultural produce aims at ensuring remunerative prices to the growers for their produce with a view to encouraging higher investment and production. Towards that end, minimum support prices for major agricultural products are announced each year, which are fixed after taking into account the recommendations of the Commission for Agricultural Costs and Prices (CACP). The CACP, while recommending prices takes into account all of the following factors:

- Cost of Production
- Changes in Input Prices
- Input/Output Price Parity
- Trends in Market Prices
- Inter-crop Price Parity
- Demand and Supply Situation
- Effect on Industrial Cost Structure
- Effect on General Price Level
- Effect on Cost of Living
- International Market Price Situation (MSP)
- Parity between Prices Paid and Prices Received by farmers (Terms of Trade).

Since liberalization several policy measures have been taken with regard to regulation & control, fiscal policy, export & import, taxation, exchange & interest rate control, export promotion and incentives to high priority industries. Food processing and agro industries have been accorded high priority with a number of important relief and incentives.

Wide-ranging fiscal policy changes have been introduced progressively. Excise & import duty rates have been reduced substantially. Many processed food items are totally exempt from excise duty. Custom duty rates have been substantially reduced on plant & equipments, as well as on raw materials and intermediates, especially for export production.

The Committee on Pricing Water (as part of the National water Policy, 1987) deals with rationalizing water rates and have suggested increase in irrigation water rates in a phased manner. The pricing of water for various uses will have to take into account the paying capacity of the users including farmers and large population below poverty line.

As for regulations & control, no industrial license is required for almost all of the food & agro processing industries except for some items like: beer, potable alcohol & wines, cane sugar, hydrogenated animal fats & oils etc. and items reserved for exclusive manufacture in the small scale sector. Items reserved for S.S.I. include pickles & chutneys, bread, confectionery (excluding chocolate, toffees and chewing-gum etc.), rapeseed, mustard, sesame & groundnut oils (except solvent extracted), ground and processed spices other than spice oil and olioresins, sweetened cashew nut products, tapioca sago and tapioca flour.

Decision-Making: Strategies, Policies and Plans

The Agricultural Development Strategy was revised in 1999, as the national strategy on sustainable agriculture and rural development (SARD). The Strategy is essentially based on the policy on food security and alleviation of hunger. A regionally differentiated strategy, based on agro climatic regional planning which takes into account agronomic, climatic and environmental conditions, will be adopted to realize the potential of growth in every region of the country. The thrust will be on ecological, sustainable use of basic resources such as land, water, and vegetation, in such a way that it serves the objectives of accelerated growth, employment and alleviation of hunger.

In the accelerated growth scenario for the Ninth Five Year Plan (1997-2002), an agricultural growth rate of 4.5% per annum is expected. Allied sectors such as horticulture including fruit and vegetables, fisheries, livestock, and dairy will continue to register greater growth during the Ninth Plan period. In the Ninth Plan, targets will be achieved through a regionally differentiated strategy based on agronomic, climatic, and environment-friendly conditions. At the macro level, the agriculture development strategy will be differentiated by broad regional characteristics of an agro-economic character. The northwestern high productivity regions will promote diversification and high value crops, and strengthen linkages with agro-processing industries and exports. The Eastern region, with abundant water, will exploit this productivity potential through flood control, drainage management, improvement of irrigation facilities, and improved input delivery systems. The water scarce peninsular region, including Rajasthan, will focus on efficient water harvesting and conservation methods and technologies based on a watershed approach and appropriate farming systems. Ecologically fragile regions, including Himalayan and desert areas, will concentrate on eco-friendly agriculture.

Animal husbandry and dairying will receive grater attention for development during the Ninth Five Year Plan as this sector plays an important role in generating employment opportunities and supplementing includes of small marginal farmers and landless laborers, especially in rain fed and drought-prone areas. Effective control of animal diseases, declaration of disease-free zones, scientific management of genetic stock resources, breeding, quality feed and fodder, extension services, enhancement of production, productivity and profitability of livestock enterprise will be given greater attention. The growth value of the output from the livestock sector is estimated to be 26% of the total value from the agricultural sector.

To make self-employment programmes more effective in the Ninth Five Year Plan, there will be a shift in strategy from an individual beneficiary approach to a group and/or cluster approach under the Integrated Rural Development Programme (IRDP). This will facilitate higher investment levels to ensure project viability. In addition, this approach will include skills development of the poor through an in built training component, upgrading of technology, establishment of forward and backward linkages, availability of appropriate infrastructure, and market tie-ups. A new initiative for social mobilization will be implemented during the Ninth Plan to create self-managed institutions for the poor. A mechanism for training social animators to assist the poor to articulate their needs and aspirations, and form their own organizations will be implemented.

Rural poverty largely exists among the landless and marginal farmers. Access to land, therefore, remains a key element of the anti-poverty strategy in rural areas. The programme of action for land reform in the Ninth Five Year Plan will include the following: detection as well as redistribution of ceiling surplus land; upgrading of land records on a regular basis; tenancy reforms to record the rights of tenants and share croppers; consolidation of holdings; prevention of the alienation of tribal lands; providing access to wastelands and common property resources to the poor on a group basis; leasing-in and leasing-out of land will be permitted within the ceiling limits; and preference to women in the distribution of ceiling surplus land and legal provisions for protecting their rights on land.

The National Development Committee (NDC) Report has highlighted the importance of social issues, which have not been addressed in quantitative terms earlier. Role of social issues and improvement of poverty in disadvantaged group of population is very important. Animal husbandry, which includes dairy, piggary, poultry, goatary and sheep farming, is the major occupation of this group of population. The above 5 farming systems should be developed on the principle of resource based planning, which includes land, water, agro-climate, labor inputs and financial capability of disadvantaged community. The livestock farming has to look into all the above facts and more importantly to economic, environmental, and social factors. Thus, the development of remunerative farming systems for improving their economic conditions and quality of life is most important in future.

Seven basic services have been identified for priority attention. Policies and programmes relating to these areas would be given a thrust in the Ninth five Year Plan. Complete coverage is expected in a time-bound manner. These services are safe drinking water, availability of primary health service facilities, universal primary education, provision of public housing assistance to all shelter less poor families, nutritional support to children, road links to all villages and habitations, and public distribution system targeted to the poor.

India's National Water Policy (NWP) was adopted in September 1987. The National Water Resources Council (NWRC) under the Chairmanship of the Prime Minister lays down the NWP, reviews development plans and advises on implementation. The Policy envisages strategies covering ground water development, water allocation priorities, drinking water, irrigation, water quality, water zoning, water conservation, flood control and management. In the context of water use, the main issues are the pricing of water for various end uses including drinking, irrigation and industrial use. The NWP of the Government of India accords highest priority to drinking water supply. The State Governments in India make their water policies within the overall framework of the NWP.

Though there has never been a single comprehensive rural energy policy for the country, the government, through it various committees such as Fuel Policy Committee (1974), Working Group on Energy Policy (1979), Advisory Board on Energy (1985), Energy Demand Screening Group (1986), etc. has formulated programmes aimed at rural energy and implemented through various ministries. The basic issues borne in mind when formulating policies have been (a) Technology choices, (b) Dissemination approach, (c) Commercialization and (d) Capacity building.

Decision-Making: Major Groups Involvement

The Panchayati Raj Institutions (PRIs) will function as effective institutions of local self-governance and they will prepare plans for economic development and social justice and implement them. The PRIs will be the umbrellas for the integration of sectoral programmes with poverty alleviation and rural development programmes. The Council for Advancement of People's Action and Rural Technology (CAPART) will continue to provide projected financial assistance to voluntary organizations, which will have to play a more dynamic role in empowering the poor through advocacy, awareness generation and formation of Self-Help Groups (SHGs) during the Ninth Plan.

In order to promote people's participation and create awareness, the practicing farmers, village youth and school dropouts are working as focal points for dissemination of information *e.g.* on low cost technology and producing plant material for conservation measures. Stress is being laid on organising SHGs to institutionalize people's participation to improve household production systems (cattle rearing, mushroom cultivation, sericulture, bee-keeping etc.)

Programmes and Projects

Major activities to implement the SARD policy are as follows:

1. Development of crops based on regionally differentiated strategy

2. Development of Horticultural crops
3. Adequate and timely delivery of core inputs
4. Integrated Pest Management
5. Greater use of bio-fertilizers and bio-technology
6. National Agricultural Technology Project
7. Rained farming and Watershed Management
8. Soil and Water Conservation
9. Animal Husbandry and dairying
10. Development of fisheries
11. Agricultural research and education
12. Development of Human resources

The major thrust of the agricultural development programmes in India is on improving the efficiency in the use of scarce natural resources, namely, land, water and energy. This can be achieved only through improved productivity in a cost-effective manner, which alone could increase the welfare of the farmers and agricultural labor. Balanced and integrated use of fertilizers, agricultural credit, institutional support, accelerated investments in agriculture, enhancing the competitiveness of agro-exports, creation of additional irrigation facilities etc. have been given encouragement through various schemes and activities of the Government of India.

A wide range of approaches have been employed to address problems of land degradation, some of which include:

- Prevention of soil loss from the catchments
- Promotion of multi-disciplinary integrated approach to catchment treatment.
- Improvement of land capability and moisture regime in the watersheds.
- Promotion of land use to match land capability
- Reduction of run-off from the catchments to reduce peak flow into the river system.

- Upgrading of skills in the planning and execution of watershed development programme.
- Increase of productivity of land affected by alkalinity for increasing sustainable agriculture production.
- Identification of critical degraded areas,
- Generation of data on land suitability and capability for regulating land use.
- Preparation of soil resource map and inventory of soil and land resources.
- Development of technical skills in soil and water conservation
- Building up and strengthening of land capability of State Land Use Boards.

Promotion and implementation of land use policy related to land base programme.

Coordination and regulation of programmes relating to land resources, conservation, management and development at State level.

Various Soil and Water Conservation Programmes have been launched in response to the need for conservation and rehabilitation of degraded land including:

- Strengthening of State Land Use Boards (SLUBS)
- National Land Use & Conservation Board (NLCB)
- Soil Conservation Training Centre DVC Hazaribagh (Plan 7 Non-Plan).
- Centrally Sponsored Scheme of Soil Conservation for Enhancing Productivity of Degraded Lands in the Catchments of River Valley Projects.
- Centrally Sponsored Scheme of Soil Conservation in the Catchments of Flood Prone Rivers
- Centrally Sponsored Scheme for Reclamation of Alkali (Usar) Soils.
- EFC Assisted Project for Reclamation and Development of Alkali land in Bihar and U.P.

- Uttar Pradesh Sodic Land Reclamation Project with World Bank assistance.
- Watershed Development Project in shifting Cultivation Areas of North Eastern States (WDPSCA).
- Indo-German Bilateral Project on Watershed Management.
- Reclamation of Marginal and shallow ravines in the states of Uttar Pradesh, Madhya Pradesh, Gujarat and Rajasthan.
- Centrally Sponsored Scheme for Reclamation of Saline Soils including Coastal Saline and Sandy Areas.
- Centrally Sponsored Scheme for Amelioration of Acid Soils.

Pests are an inevitable part of agriculture. So it is equally inevitable that humans have been attempting to find ways of reducing the pests¡¯ share of their crops. The threat posed by these pests has been perceived to be so great over the last 60-70 years that a process called the integrated pest management (IPM) has been developed in the name of crop protection. All three components including the pesticide promoters, the pesticide antagonists and the fence-sitting demanders of knowledge support IPM.

To alleviate the ill effects of pesticides, India has officially adopted IPM as its policy and is a prominent feature in recent Five Year Plans. In fact, among other nations in Asia, India was first to adopt the policy. One of the manifestations of this policy is the Central IPM Centre (CIPMC), of which there exists at least one in each state. Their functions include crop surveys, training the trainers of ¡®IPM farmers¡¯, and rearing natural control agents. Central efforts on plant protection are being targeted to popularise environment friendly IPM approach. Greater relevance is given to bio-control of pests under the IPM and human resources development.

On a broader scale, IPM is defined and explained in terms that encompass the farm families & their environment, and regional food security. The essential element for IPM includes one or more management activities that are carried out by farmers that result in the density of potential pest populations being maintained below levels at which they become pests, without endangering the productivity and profitability of the farming system as a whole, the health of the farm family and its livestock, and the quality of the adjacent and downstream environments.

Major steps towards safe and appropriate use of pesticides include:

- Promotion of Integrated Pest Management
- Implementation of Insecticides Act
- Training in Plant Protection
- Locust Control & Research
- Strengthening and Modernization of Plant Quarantine Facilities in India

Since 1992, 4391 farmers have been trained in IPM (information from States of West Bengal, Orissa, Kerala and Karnataka awaited). Crop types covered with IPM programmes are: package for 5 Kharif crops and 14 Rabi crops. Although many farmers practice IPM in India, the focus is mainly on rice. There is considerable scope to extend the movement to other crops grown in paddy fields and in non-irrigated and upland areas; in particular vegetables, groundnut, pulses, sugarcane, castor sunflower and other oilseeds, sorghum and cotton.

The benefits of IPM programmes include

- Effective control of pests and diseases by emphasising the need based application of pesticides. Prevention of indiscriminate use of pesticides;
- Monitoring /Forewarning of pest disease situation;
- Promotion of biocontrol agents (including field releases of laboratory reared agents) and Pest surveillance and monitoring vis-a-vis biocontrol agents; and
- Conservation of environment and ecosystem.

In India the average annual precipitation is nearly 4000 cubic km (km^3) and the average flow in the river systems is estimated to be 1869 km^3. Because of concentration of rains only in the 3 Monsoon months, the utilizable quantum of water is about 690 km^3. Quantum of ground water extracted annually is-about 432 km^3. Thus, on an average, 1122 km^3 water is available for exploitation and is considered adequate to meet all the needs. However, the situation is complicated because this water is not uniformly available either spatially or temporally. Six of the 20 major river basins in India suffer from water scarcity. Water has already become one of the most limiting resources in the country. Solving scarcity of water both in quantity and quality, national programmes (Preventive & Mitigative Action Plans) have been launched to tackle the situation which include:

1. *Guidelines for Ground Water extraction and use.*

Contribution of ground water for irrigation as well as industrial use and drinking has been on the increase during the last two decades. Indiscriminate extraction of ground water already poses the threat of aquifers going dry in some parts of the country. The Central and State Ground Water Boards have, therefore, prepared Ground Water Availability Maps and prescribed extraction rates in a bid to ensure that extraction is balanced with recharge. The country has been ZONED depending upon whether water is available in plenty, or it has already become scarce in the region. Accurate determination of ground water reserves can be done through actual Bore Hole Data in a given region. Extraction of ground water is prohibited in some regions where water depletion has already become critical.

2. *Management of Lakes*

Natural and man-made lakes happen to be a major source of water supply in many regions in India. Excessive siltation, variation in run-off and changing land use in the watersheds has contributed to depletion of these water bodies. The water quality in lakes is also affected by run-off loaded with fertilizers, insecticides, pesticides coupled with discharges from industries as well as human settlements. Major interventions for improving the lake systems in the country include Watershed Management, Dredging operations, emphasis on treatment of effluents before discharge into the lakes and disposal of solid wastes away from the shores of the lakes.

Water use efficiency is presently estimated to be only 38 to 40% for canal irrigation and about 60% for ground water irrigation schemes. India's per capita water availability per year (1991 census) was estimated at 2209 cubic metres against the global average of 9231 cubic metres. In the total water use in 1990, the share of agriculture was 83%, followed by domestic use (4.5%), industrial use (2.7%) and energy (3.5%). The remaining 6 per cent were for other uses including environmental requirements.

The projected total water demand by the year 2025 is around 1050 cubic kilometres against the country's utilisable water resources of 1132 cubic kilometres. The share of agriculture in total water demand by the year 2025 is expected to be about 74 to 75 per cent. Irrigation, being the major water user, its share in the total demand is bound to decrease from the present 83% to 74% due to more pressing and competing demands from other sectors by 2025 A.D. It is estimated that a 10% increase in the present level of water use efficiency in irrigation projects, an additional 14 m. ha area can be brought under irrigation from the existing irrigation capacities which would involve a very moderate investment as compared to the investment that would be required for creating equivalent potential through new schemes. Thus, the need to improve the present level of water use efficiency in general and for irrigation in particular assumes considerable significance in perspective water resource planning.

In order to promote the process of improvement in water management through upgrading of the main systems of selected irrigation schemes the National Water Management Project (NWMP), an externally aided project (EAP) was implemented during the period 1987-95. Now, the Ministry of Water Resources has initiated follow-up action on NWMP-II with an estimated cost of Rs.2880 crore for 7 years. In more recent times, the Water Resource Consolidation Project (WRCP) has been taken up in the States of Haryana, Orissa and Tamil Nadu, which envisages the completion of some major and medium irrigation projects and strengthening of institutions through Participatory Irrigation Management/Irrigation Management Transfer (PIM/IMT).

It is a fact that water logging has been observed in some of the irrigated commands and the same is adversely affecting the productivity in these areas. Integrated and coordinated development of surface and ground water is widely recognized as a most suitable strategy for irrigation development in alluvial plains. Gradual rise in water table and related problems of water logging and soil salinity/alkalinity have surfaced mainly because of the lack of drainage provision, improper waste management, inadequate maintenance etc.

Conjunctive use of surface and ground water will not only increase the irrigation potential, but also mitigate the problem of water logging. The technologies of irrigation from both surface and ground water may be integrated in a complementary manner, in order to achieve sustainable optimum agricultural production and equity. Such integration may be brought about in one or more of the following ways:

- *conjunctive in space:* Some parts of the command may be irrigated exclusively by surface water and other parts by ground water.
- *conjunctive in time:* Parts of the command may be irrigated alternatively by surface and ground water at one time of the growing period/crop season.

- ***conjunctive by augmentation:*** Supplies from one source are augmented by those from other sources *e.g.* augmentation tube wells.

Considering the problem, reclamation of waterlogged areas has been included as a new component of CAD Programme since 01.04.1996. Ministry of Water Resources has organised two Workshops on the subject and held many training programmes to create awareness among functionaries and farmers. The Ministry has also constituted and Coordination Committee under the Chairmanship of Additional Secretary to look into the problem. A manual has also been developed to give technical input to States to identify the problem areas and take up preventive and remedial measures suitably. A total target of 60,000 ha. has been kept for reclamation of waterlogged areas during the Ninth Plan. The efforts of the Ministry got response from the States and they have identified the schemes, which have further been posed to the Ministry for concurrence. The Ministry has given administrative approval to 129 schemes so far during 1998-99 and 1999-2000 covering an area of 39325.46 ha. The work has been taken up by the States and is likely to gain momentum to achieve the target of 60,000 ha during Ninth Plan.

Status

Land, which is the most precious heritage and physical base of bio-mass production of life supporting systems is finite, and thus a non-renewable endowment. India's share of land is fixed at about 329 m. ha., which is heterogeneous in different parts and regions of the country with a definite set up, capabilities and suitability for different land resources. Conservation of land resources can promote sound land use to match with the land capabilities or suitability and to initiate correct land resources, development/ suitability in the country.

The agriculture sector has a vital place in the economic development of India as it contributes 29.4% of GDP and employs about 64% of the workforce. Significant strides towards ensuring food security have been made in agriculture production. Food grain production registered an annual growth rate of 3% from 1984-85 to 1994-95. The significant improvement in agriculture productivity has helped reduce rural poverty. Though capital formation in agriculture grew at the rate of 6.05% during 1989-90 to 1994-95, its share in the total gross capital formation declined to 10.85% from 18.86% in 1980-81 (using 1980-81 prices).

Food grain production increased from 168.4 million tonnes in 1991 to an expected level of 196.0 million tonnes in 1997, the terminal year of the Eighth Five Year Plan. Lack of any significant breakthrough in seed technology is perhaps one of the main reasons for the slow growth in good grain output during the nineties. The production of commercial crops like sugarcane 9,283 million tonnes), oilseeds (22.4 million tonnes), cotton (13.1 million bales) was at a record level in 1995-96. The organised upland tea and coffee plantations, the extensive and often dense coastal strips of coconut trees, and the subterranean tuber and root cops characterize the variegated nature of the horticultural potential in the country. The production of flowers has emerged as a promising area of high growth in recent years, particularly for its export potential. However, due to lack of technology and poor infrastructure support for handling, packing, processing and preservation, substantial post harvest losses of fruits and vegetables still characterize the horticulture sector.

The country's irrigation potential was 89.56 million ha by the end of 1996-97; comprising 32.96 million ha under major and medium projects, and 56.60 million ha under minor irrigation schemes. The domestic production of fertilizers falls short of requirements. Integrated Pest Management (IPM) in India includes pest-monitoring promotion of biological control of pests, and organising demonstrations and training.

Animal Husbandry is an important source of self-employment and subsidiary occupation in rural and semi-urban areas, especially for people living in drought prone, hilly, tribal and other poorly developed areas, where crop production along may not fully sustain them. Exported agricultural products include food grains, tobacco, cashews, groundnuts, beverages, sugar, molasses, horticulture and floriculture products, processed fruits and juices, and meat preparations. India's share in the world trade in agricultural commodities is about 1%. Agricultural exports have received special attention from the Government because of the potential for raising farm incomes, tackling unemployment, and earning foreign exchanges. A number of policy changes have been introduced to give an impetus to agricultural exports.

Agriculture is now reckoned to be the largest consumer of water, accounting for some 80% of total water use. To maximize food supply for humanity, land irrigation has been practiced for centuries. Irrigation plays a large role in increasing arable production and cattle-breeding efficiency, with irrigated farming expected to continue to develop intensively in the future. Thus, irrigation has now become the principal water user. The irrigation potential was 22.6 million ha in 1951 with food production of 50 million tons. The food production has quadrupled now to about 200 million tons due to four-fold increase in irrigation potential at over 10 million ha. As recently reassessed by the Ministry of Water Resources the country's ultimate irrigation potential is tentatively estimated at 139.89 m. hectares, comprising 58.46 m. hectares of major & medium irrigation and 81.43 m. hectares of minor irrigation as against pre-revised ultimate irrigation potential of 113.50 m. ha. The full development of ultimate irrigation potential by construction of major, medium and minor irrigation projects by 2025 would be necessary to meet the food requirement of the projected population.

India is the second most populous nation in the world. 70% of the population of India, which is close to 700 million, still live in the rural areas. Meeting their energy requirements in a sustainable manner continues to be a major challenge for the country. Almost 75% of the total rural energy consumption is in the domestic sector. For meeting their cooking energy requirements, villagers depend predominantly on biomass fuels such as wood, animal dung and agricultural residues, often burnt in inefficient traditional cooking stoves. The main fuels used for lighting in the rural households are kerosene and electricity. Irrigation is mainly through electrical and diesel pump sets, while the rural industries and the transport sectors rely primarily on animal power and to some extent on commercial sources of energy like diesel and electricity.

Of the total energy consumption in the country, almost 60% is met by conventional energy sources and the rest is met by non-conventional and renewable energy sources. This energy use pattern has serious implications both on the environment as a whole as well as on the users. Fuel wood requirements have contributed to the degradation of forests. Degradation of forests has associated implications regarding CO_2 sequestration. Further, this has led to villagers, especially women and children traveling longer distances and spending more time in collecting fuel wood, switching to inferior, fuels, and even altering food habits to reduce fuel consumption affecting the nutrition levels. Given the exploitation processes of natural resources, this situation is likely to worsen in the years to come. Rural energy systems are further strained by the inability of people to shift to commercial fuels like electricity, LPG and kerosene because of low purchasing powers and limited availability. The large subsidies on electricity for agriculture and kerosene have also been a cause of concern for energy planners.

To redress these problems, several efforts have been made both by governmental organizations and non-governmental organizations in the form of national programmes for rural electrification, and promoting renewable energy technologies like biogas, improved cooking stoves and solar cookers. However, in spite of the existence of these programmes for nearly two decades, their impact on the rural energy scenario has been limited. Over the last few years, in line with economic liberalization, there have been efforts towards bringing about commercialization implemented in the past two decades in order to formulate a meaningful rural energy policy at the national level.

The present supply-demand scenario indicates that biomass would continue to be the mainstay of the rural energy sector in the foreseeable future. The penetration of various commercial fuels will remain quite low, and at the present rate, it would take a long time for the RETs (Renewable Energy Technology) to make any significant impact on the sector. Therefore, any policy formulated to deal with rural energy will have to look for highly innovative options and make judicious investment choices.

The demand of pesticides for the year 1998-99 has been estimated to be around 57, 240 million tonnes. The overall availability of pesticides in India is satisfactory. The consumption of chemical fertilizers during the year 1997-98 was 161.88 lakh tons of nutrients. The consumption of nitrogen, phosphate and potash in fertilizers increased by 5.8%, 31.5% and 33.3% respectively, from 1997 to 98. During the same year the consumption of both Di-Ammonium Phosphate and Muriate of Potash were 53.76 and 17.29 lakh tonnes respectively. All chemical fertilizers except urea continue to be decontrolled.

Challenges

It is estimated that about an average 16.75 to/ha/year of soil are lost through erosion every year in India *i.e.* more than 5,000 million tons of topsoil is eroded annually. A close look at the present health of the soil and water resources reveals their wanton misuse and degraded environment.

Almost 173.64 m. ha. covering slightly half of the country, are threatened by various types of degradation such as salinity, alkalinity, water logged areas, ravinous and gullied lands, areas under ravages of shifting cultivation, desertification, etc. About 800 hectares of arable land are being lost annually due to ingress of ravines. There are specific problems of land degradation due to open cast mining operations using good productive land for brick kilns coastal erosion and seawater ingress, excessive erosion and land slides in the crumbling hill areas. Our forests and grasslands have been over exploited. Frequent occurrences of floods and droughts in different parts of the country are evidence of improper land use in the catchments and inadequate conservation of rainwater. The problem of land degradation has brought us face to face with the ever increasing depletion of the productivity and the basic land stock through nutrient deficiencies on the one hand and the ever growing demand for food, fodder, fiber, fuel, land based industrial raw materials and may non-farm land uses on the other hand.

Over the last two decades, there has been a considerable decline in the incidence of rural poverty. However, a large number of persons continue to live below the poverty line. Hence, there is a need for continued direct State intervention for the eradication of poverty. While the programmes for self-employment and supplementary wage employment would continue in the Ninth Five Year Plan, these would be redesigned to make them more effective as poverty alleviation instruments. They will also be integrated with various sectoral and area development programmes within the umbrella of the Panchayati Raj Institutions (PRIs).

There is urgent need to reduce the dependence on fertilizer imports by improving output and productivity in fertilizer production units. Improvements in energy efficiency in the fertilizer sector to reduce the cost of fertilizer production are significant. The promotion of a higher seed replacement rate will be emphasized. In the post General Agreement on Tariffs and Trade (GATT) period, new plant variety protection rights make it necessary to augment facilities for the registration of varieties.

Capacity-building, Education, Training and Awareness-raising

Increased mechanization in agriculture has created demand for more trained manpower for the operation, maintenance, and management of agricultural machinery. The Government has set up Farm Machinery Training and Attesting Institutes to provide better quality equipment to farmers. The Indian Council of Agricultural Research (ICAR) plays a crucial role in promoting science and technology and its application in agriculture. A National Gene Bank, which is the biggest in Asia, has been opened in New Delhi.

The Training for Rural Youth for Self-Employment (TRYSEM) will be revamped in its design, curriculum and method of training in order to improve the employment opportunities of the poor. It will focus on activities in which the rural youth are already engaged and where there exists a potential for skill upgrading or else on activities, which would enhance production under group-cluster approach. The artisans in rural areas, despite their rich heritage and skills, belong to the poverty group. The existing programme aimed at upgrading their skills and improving their production capabilities, by supplying them with modern tool kits, would be strengthened and expanded in the Ninth Plan. This would facilitate enhancing the productivity and income levels of the rural artisans.

The Development of Women and Children in Rural Areas (DWCRA), which is based on a group approach, has been successful in empowering women and in improving their economic status in selected States. A mechanism for replicating the successful DWCRA groups will be sought. Thrift will be the starting point for the formation of SHGs. A greater integration of DWCRA with IRDP and ATRYSEM will be attempted to provide women's groups with greater access to financial resources and training.

Human resource development in plant protection and various disciplines of pesticides is being achieved by organising regular and short term training programmes at National Plant Protection Training Institute, Hyderabad.

Information

National information on sustainable agriculture is made available to decision-makers, advisory organizations and farmers via the Internet: *http://goidirectory.nic.in*. The government of India has initiated the following activities with regard to analysis and collection of information on various production systems and technologies.

- Development of on-farm and off-farm programmes to collect and record indigenous knowledge
- Analysis of the overall effects of technological innovations and incentives on farm household income

The I.T. Division, Ministry of Agriculture, GOI provides world-class services in terms of information and communication relating to agriculture nationally and internationally. Networking of information right from the level of farmers and the village/ block to district headquarters on the one hand and the Central Government Departments and the Attached & Subordinate Offices as well as other autonomous organisations, non-Government organisations etc. on the other is being established for maximising efficiency and productivity. This is being done by facilitating availability of information with speed, quality and economy in every area connected with agricultural productivity e.g. fertiliser, insecticides pest attack, drought and other natural disasters, marketing, storage, pricing etc. The effort will be to bring about qualitative change in management of agriculture through information management with the help of the latest information technology. A programme to support early warning systems for monitoring food supply and other associated factors in both urban and rural areas, a programme has been launched to Strengthen the Information Technology apparatus in the Department of Agriculture and Cooperation.

The objectives of the programme is to provide I.T. tools to officers in agriculture departments in their day-to-day working and to have a National Agriculture Informatic Centre as reservoir of all data relating to agriculture and to set up an Early Warning Information System for crop monitoring and forecasting on day-to-day basis. The scheme provides for:

- Total computerisation in the headquarters through Local Area Network (LAN)
- To have separate I.T. Division to monitor and formulate I.T. schemes of the department
- To create historical data base
- To disseminate information through workshops/ seminars etc. on IT related activities
- To establish live and independent network for information gathering from field to headquarters
- To co-relate and analyse the field information and available data and make forecasting on crop prospects on day-to-day basis.

Research and Technologies

Research efforts will be accelerated through biotechnology, micro-biology, genetic improvement of crops including hybrid technology, genetic up gradation of animal harvest technology, etc. In agricultural education, the thrust will be on human resource development through upgraded teaching facilities. The existing infrastructure for technology transfer will be made more effective and responsive to meet farmers needs.

Farm Mechanisation programmes of the Government have been directed towards selective mechanisation with the aim of optimum utilisation of the available sources of farm power. The programmes emphasises popularisation of improved and modern agricultural implements and machines through various Schemes. Farmers have been provided financial assistance for owning tractors and other improved agricultural implements and machines. The infrastructure for human resource development, and for testing and evaluation of agricultural implements/machines has been established. The emphasis has also been laid on the safety of farmers in operation of agricultural machines. The programmes have resulted in the increased adoption of improved farm machines and equipment by the farmers.

The Ninth Plan programmes give a special thrust to a sustainable and all-round agricultural development in the country through a pragmatic farm mechanisation strategy for the different agro-climatic zones of the country. The main features of different schemes of agricultural implements and Machinery Division are given below:

- Strengthening of Farm Machinery Training and Testing Institutes (FMTTIs) in Madhya Pradesh, Haryana, Andhra Pradesh and Assam.
- Promotion of Agricultural Mechanisation.
- Development of Industrial Designs of Agricultural Implements including Horticultural equipment and their trials at Farmers' field.
- Conducting Study and Formulating Long Term Mechanisation Strategy for Each Agro-Climatic Zone.
- Promotion of Agricultural Equipments in North-Eastern States.
- Establishment of Farm Machinery Training & Testing Institute at Tamil Nadu.
- Comprehensive Scheme on Modernisation of Agriculture through Mechanisation.

To supplement the efforts of State Governments for increasing the production and productivity, six Centrally-Sponsored and one Center-sector schemes are being implemented by the Crops Division of the Ministry of Agriculture in different States. Under these schemes, emphasis is being laid on the transfer of improved crop production technologies through organization of field demonstrations on farmer holdings and farmer trainings. Additionally, to motivate the farmers to adopt the improved crop production technologies incentives are being provided through the respective schemes on the use of inputs like certified seeds/quality seeds, improved farm implements, sprinkler/drip irrigation system etc. Such programmes include:

- Integrated Cereals Development Programme in Rice Based Cropping Systems Areas (ICDP-Rice)
- Integrated Cereals Development Programme in Wheat Based Cropping Systems Areas (ICDP-Wheat)
- Integrated Cereals Development Programme in Coarse Cereals Based Cropping Systems Areas (ICDP-Coarse Cereals)
- Sustainable Development of Sugarcane Based Cropping System (SUBACS)
- Special Jute Development Programme (SJDP)
- Intensive Cotton Development Programme
- Mini-kit Demonstration programme of Wheat, Rice and Coarse Cereals including propagation of new Technologies.

Over the last 12000 years of evolution of agriculture practices, the Science & Technology inputs have only succeeded in evolving just about 10% of the genetic stock found in the wild into palatable and higher yielding cereals, fruits and vegetables. Food security demands that the remaining 90% of the stock should be preserved firstly, for developing additional higher yielding varieties to feed the increasing population, and secondly, to protect and provide immunity to the existing higher yielding varieties when they are under attack from insects, pests and epidemics.

Realising the importance of Genetic Stock for food security, the National Bureau of Plant Genetic Resources of ICAR has long been identifying areas rich in bio-diversity and Gene Pool for cereals, fruits and vegetables. We need to protect our rich heritage of herbs, shrubs & medicinal plants. Bio-Banks created have seed as well as tissue samples of the requisite crops. In light of the International Convention on Bio-diversity to which India is a signatory, "Recombinant DNA safety Guidelines" for personnel and environmental safety in the use of genetically manipulated organisms in research, manufacture and applications have been evolved. Declaration of eco-sensitive zones, introduction of the National Biodiversity Strategic Action Plan (BSAP) and other initiatives are further steps towards conservation and sustainable use of biological resources.

Financing

The funding of major programmes are done mainly through the national budget. A public sector outlay of Rs. 42642 crore (at 1996-96 prices) has been earmarked for development of agriculture and allied activities in the Ninth Plan.

The emphasis on agricultural credit has continued to be on progressive institutionalization for providing timely and adequate credit support to farmers with particular focus on small and marginal farmers and weaker sections of society for increasing agricultural production and productivity. The Government of India has taken many policy initiatives for strengthening the rural credit delivery system to support the growing credit needs of the agricultural and rural sectors. The Policy essentially laid emphasis on augmenting credit flow at the ground level through credit planning, adoption of region-specific strategies and rationalization of lending policies and procedures to enable the farmers to adopt modern technology and improved agricultural practices.

- Agricultural credit is disbursed through multi-agency network consisting of Commercial Banks (CBs), Regional and Rural Banks (RRBs) and Cooperatives.
- The Cooperative Credit Institutions, both the short and long term structures, have emerged over the years as the Prime institution agencies for dispensation of rural credit.

In terms of network, coverage and outreach. Cooperatives have sizeable presence and play a significant role in meeting the short-term requirements of agriculture. However, several developments over a period of time have left the Cooperative Credit Structure (CCS) facing severe problems, which have restricted their ability to function viably and perform effectively the task of reaching out to all segments of farming community and meet in full their requirements of credit. In order to build up a strong and viable CCS, a proposal for revamping of CCS is under active consideration of Government of India.

- To provide adequate and timely support from the banking system to the farmers for their cultivation needs including purchase of all inputs in a flexible and cost effective manner, a model Kisan Credit Card Scheme was formulated and introduced in the year 1998 for implementation by all the rural financial institutions in the country. Under the Scheme, banks may provide the Kisan Credit Cards to the farmers who are eligible for sanction of production of credit of Rs. 5,000/- and above and there is no upper limit. The credit extended under the scheme is in the nature of revolving Cash Credit and provide for any number of draws and repayments within the limit.
- A Scheme for providing insurance cover to farmers, known as comprehensive Crop Insurance Scheme was in vogue in the country since 1985. To enlarge the coverage in terms of farmers (loanee and non-loanee both) crops and risk under Crop Insurance, the Government of India have decided to implement a new Crop Insurance Scheme entitled "National Agricultural Insurance Scheme" (Rashtriya Krishi Bima Yojna) from Rabi 1999-2000. The Scheme provides financial support to the farmers in the event of failure of their crops due to all types of natural disaster as well as pest attacks and diseases.
- Rural Infrastructure Development Fund (RIDF) was

created in the year, 1995-96 to boost public investment in development of rural infrastructure. The assistance provided under RIDF has primarily facilitated augmenting resources of the State Governments for investment in rural infrastructure projects including projects that remained incomplete for want of resources. This, in turn, accelerated creation of employment opportunities and production base in the rural areas. The fund under RIDF is made up of contributions from the Indian Scheduled Commercial Banks against their shortfall in agriculture target lending up to an extent of 1.5 percent of net bank credit.

Cooperation

India is a signatory to several International Conventions like CITES, International Whaling Convention (IWC); Convention on Migratory Species (CMS) and the World Heritage Convention (WHC).

Establishment of a Protected Areas Network, under the Wildlife (Protection) Act, 1972, comprising of Biosphere Reserves, National Parks and Sanctuaries, both terrestrial and aquatic, has been a positive step towards conservation of animal genetic resources. This Network today comprises of 10 Biosphere Reserves, 89 National Parks, 504 Sanctuaries, along with dedicated conservation programmes such as Project Tiger, Crocodile Rehabilitation and project Elephant. India has recently taken the lead in the formation of the Global Tiger Forum. The Central Zoo Authority caters to the ex-situ conservation of wildlife through 275 zoos, deer parks, safari parks and aquaria, etc.

Effective measures for control of illegal trade in wildlife and its products at national and international level, both through the States/UTs as well as Regional Offices of Wildlife Preservation under Ministry of Environment and Forests have been taken. Wildlife Week was celebrated from 2nd to 8th October 1998 all over India. Various functions for generating awareness about wildlife conservation were held by the State/UT Governments. During the week, essay competitions, debates, clay modeling, free trips to national parks and sanctuaries, drawing competitions etc. were organised. The Wildlife Institute of India (WII) organised three courses and 46 officers and students were trained during the year. Efforts to build-up professional skills in Protected Area Management through training of professional managers for protected areas through training of professional cadre in all aspects of wildlife are continuing the WII.

This information was provided by the government of India to the 5th and 8th Sessions of the United Nations Commission on Sustainable Development. Last Update: February 2000.

ATMOSPHERE

Decision-Making: Coordinating Bodies

Environmental problems and issues received special attention of the Government of India during the beginning of the Fourth Five Year Plan. As a follow-up step, a National Committee of Environmental Planning and Co-ordination (NCEPC) was set up in 1972 under the Department of Science and Technology. A separate Empowered Committee was set up in 1980 for reviewing the existing legislative measures and administrative machinery for ensuring environmental protection and for recommending ways to strengthen them.

On the recommendations of this Empowered Committee, a separate Department of Environment was set up in 1980 which was subsequently upgraded to a full-fledged Ministry of Environment and Forests in 1985 to serve as the focal point in the administrative structure of the Government of India for the planning, promotion and co-ordination of environmental and forestry programmes. Other Government partners in carrying out environmental protection activities include:

- The State departments of environment;
- Central and State Pollution Control Boards;

- the Botanical and Zoological Survey of India;
- the Forest Survey of India;
- the National River Conservation Authority (formerly Central Ganga Authority);
- the National Afforestation and Eco-development Board;
- the Indian Council for Forestry Research and Education; and
- the Wildlife Institute of India and the National Museum for Natural History.

The Government of India has entrusted the work relating to Ozone layer Protection and phase out of the Ozone depleting substances programme under the Montreal Protocol to the Ministry of Environment and Forests.

There are several inter ministerial committees and working groups to coordinate issues relating to policy and legislation. In case of implementation of the Montreal Protocol the Empowered Steering Committee consists of members from the Ministry of Science and Technology and other line Ministries. The Ministry of Environment and Forests has set up an Ozone Cell as a national unit to implement the Protocol and the Ozone Depleting Substances phase out programme in India.

The decisions related to Ozone Depleting Substances phase-out and protection of Ozone Layer are being taken at the Central Government level. The Government of India frequently consults NGO's research organizations and regional expertise as and when needed.

Decision-Making: Legislation and Regulations

India's approach towards implementation of the UNFCC and associated environmental parameters is covered within policy declarations:

the National Conservation Strategy and policy statement on Environment and development (1992) and the policy statement on abatement of pollution (1992) for regulating various environmental parameters;

- In addition various other enactments such as Air pollution (prevention and control) Act 1981 amended in 1987 and Motor Vehicles Act 1939, amended in 1988;
- the Forest Conservation Act amended in 1988 contribute significantly towards minimizing the causes of climate change.
- The Environmental Protection Act 1986 is an umbrella legislation and it also empowers the government to formulate statutory rules for fulfilling various requirements.
- Further EIA has been made statutory for various developmental activities and the Coastal Resource Zones notification (1991) provides guidelines for protection and management of coastal zones.

India has taken a series of fiscal and regulatory measures to facilitate Ozone Deleting Substances (ODS) phase out in the country. Trade in ODS with non-Parties to the Montreal Protocol has been banned. Harmonized classifications of commodity codes consistent with the international system have been introduced. Controlled substances under the Protocol have been brought under the ambit of licenses for the purpose of export and import. The Ozone Depleting Substances (Regulation and Control) Rules, 2000 under the Environment Protection Act 1986 have been notified to Control and phase out production and consumption of Ozone Depleting Substances in order to comply with the Montreal Protocol.

The Government decided in 1995 to fully exempt payment of Customs and Excise Duties on capital goods required to implement ODS phase out projects funded by the multi lateral funding (MLF). This benefit was later extended for all MLF eligible projects whether or not MLF assistance was requested/available at the time of implementation of ODS phase out project. The Reserve Bank of India advised all commercial banks in September

1997 not to finance/refinance new investments with ODS technologies.

Decision-Making: Strategies, Policies and Plans

The following is an overview of the strategy for prevention and control of pollution:-

The policy statement on Abatement of Pollution, adopted in 1992, provides instruments in the form of legislation and regulation, fiscal incentives, voluntary agreements, educational programmes and information campaigns to prevent and control pollution of water, air and land. Since the adoption of the policy statement, the focus of activities has been on issues such as promotion of clean and low waste technologies, waste water minimization, reuse/recycling, improvement of water quality, environment audit, natural resource accounting, development of mass-based standards as well as institutional and human resource development. The issue of pollution prevention and control entails a combination of command and control methods; voluntary regulatory and fiscal measures; promotion of awareness and involvement of public.

To facilitate industries in preparing environmental statements, sector-specific environmental audit manuals have been prepared. A software package, Paryavaran, along with a user manual has been prepared for analysis of information submitted in these environmental statements and distributed to all the State Pollution Control Boards (SPCB). Training programmes are also being organized for officials of Central Pollution Control Boards (CPCB) and SPCB to enable them to use the software.

An "Eco-mark" label has been introduced to label consumer products that are environment-friendly. Nineteen products have been identified for labeling and 18 notifications have been issued so far on different products criteria. The Bureau of India Standards (BIS)/Directorate of Marketing and Inspection (DMI) is the implementing agency for this scheme. So far one license has been granted by the BIS to a product under the soaps category. Under the scheme for adoption of clean technology in small-scale industries and for extending necessary technical support, training and awareness programmes for personnel in Small Industry Development Organization and for entrepreneurs are being organized. 'From Waste to Profits', a manual giving guidelines for waste minimization, has been prepared. Sector-specific manuals on waste minimization in the areas of pulp and paper, pesticides formulations and textiles, dyeing and printing and electroplating, have also been prepared.

Waste Minimization Circles (WMCs) are being established to promote group efforts in increasing productivity and improving the environmental conditions in small and medium-scale industries through adoption of waste minimization techniques. Fifteen Waste Minimisation Circles have been established so far in different industrial clusters across the country.

An Indian Centre for the Promotion of Cleaner Technologies (ICPC) with a network of institutions including industries, academic institutions and other user agencies is being set up for which the World Bank has provided $2 million as grant-in-aid. The Centre will provide evaluated and ranked technology options to entrepreneurs.

Indian Forestry policy

India is one of the few countries which have a forest policy since 1894. It was revised in 1952 and again in 1988. The main plank of the Forestry Policy of 1988 is protection, conservation and development of forest. Its aims are:

(i) maintenance of environmental stability through preservation and restoration of ecological balance;

(ii) conservation of natural heritage;

(iii) check soil erosion and denudation in catchments area of rivers, lakes and reservoirs;

(iv) check extension of sand dunes in desert areas of Rajasthan and along coastal tracts;

(v) substantial increase in forest tree cover through afforestation and social forestry programmes;

(vi) steps to meet requirements of fuel wood, fodder, minor forest produce and soil timber of rural and tribal populations; (vii) increase in productivity of forest to meet the national needs;

(viii) encouragement of efficient utilization of forest produce and optimum substitution of wood and

(ix) steps to create massive people's movement with involvement of women to achieve the objectives and minimize pressure on existing forests.

The entire gamut of forest activities are being given a new orientation in the light of the National Forest Policy of 1988. In order to operationalise the National Forest Policy 1988, a National Forestry Action Programme (NFAP) is being prepared. As a part of this exercise State Forestry Action Programmes are also being prepared for each State.

Under the provisions of the Forest (Conservation) Act, 1980, prior approval of the Central government is required for diversion of forest lands for non-forest purposes. Since the enactment of the Act, the rate of diversion of forest land has come down to around 25,000 hectares per annum from 0.143 million hectares per annum, before 1980. During 1998, 851 proposals from various State and UT governments were processed under this Act.

A scheme titled 'Association of Scheduled Tribes and Rural Poor in Regeneration of Degraded Forests on Usufruct Sharing Basis' is under implementation in nine States of the country. Besides improving the forest cover, the scheme also aims at providing wage employment and usufructs to the tribal people. Joint Forest Management (JFM) is being practiced in 21 States of the country. About 7 million hectares of degraded forests in the country are being managed and protected through approximately 35,000 village Forest Protection Committees.

CENTRAL POLLUTION CONTROL BOARD

The Central Pollution Control Board (CPCB) is the national apex body for assessment, monitoring and control of water and air pollution. The executive responsibilities for enforcement of the Acts for Prevention and Control of Pollution of Water (1974) and Air (1981) and also of the Water (Cess) Act, 1977 are carried out through the Board. The CPCB advises the Central Government in all matters concerning the prevention and control of air, water and noise pollution and provides technical services to the Ministry of Environment and forests for implementing the provisions of the Environment (Protection) Act, 1986. Under this Act, effluent and emission standards in respect of 61 categories of industries have been notified.

Seventeen categories of heavily polluting industries have been identified. They are: cement, thermal power plant, distilleries, sugar, fertilizer, integrated iron and steel, oil refineries, pulp and paper, petrochemicals, pesticides, tanneries, basic drugs and pharmaceuticals, dye and dye intermediates, caustic soda, zinc smelter, copper smelter and aluminum smelter. Out of a total of 1,551 units identified under these 17 categories, 1,266 units have installed adequate facilities for pollution control and 130 units have been closed down.

The Central Pollution Control Board, in consultation with State Pollution Control Boards, has identified critically polluted areas in the country which need special attention for control of pollution. These are: Vapi (Gujarat), Singrauli (Uttar Pradesh), Korba, Ratlam, Nagda (Madhya Pradesh), Digboi (Assam), Talcher (Orissa), Bhadravati (Karnataka), Howrah (West Bengal), Dhanbad (Bihar), Pali and Jodhpur (Rajasthan), Manali and North Arcot (Tamil Nadu), Visakhapatnam and Patancheru, (Andhra Pradesh), Chembur (Maharashtra), Najafgarh (Delhi), Govindgarh (Punjab), Udyog Mandal (Kerala) and Parwanoo and Kala Amb (Himachal Pradesh).

The CPCB in collaboration with the SPCBs monitors the quality of fresh water resources of the country through a network of 480 monitoring stations located all over the country. Based on such monitoring, 13 heavily polluted and 26 medium-polluted river stretches have been identified. Under the National Ambient Air Quality Monitoring programme, 290 stations covering over 90 towns/cities monitor the ambient air quality of the country.

The Central and State Pollution Control Boards regularly conduct surveys in different cities of the country pertaining to vehicular and noise pollution, sanitation status, status of solid waste, etc. A survey on the status of solid waste conducted in 299 Class I cities of the country indicates that 62 per cent of the total solid waste generated in the country comes from the 23 metro cities of the country. The average per capita generation of solid waste for Class I cities is about 376 gms per person per day.

A total of 1,532 grossly polluting industries in 24 States/ Union Territories have been identified under the National River Action Plan. Comprehensive River Basin Documents for the rivers Ulhas, Brahmaputra, Pennar, Indus Part II, Rishkulya and Chaliyar are being prepared by the Board. The Central Pollution Control Board has a NGO Cell for interacting with NGOs. Simple water-testing kits are distributed free of cost to selected NGOs and financial assistance provided to them for conducting mass awareness programmes relating to prevention and control of pollution.

The White Paper on status of pollution in Delhi with an Action Plan for its control prepared earlier is being implemented. The Action Plan contains specific measures for control of pollution relating to vehicular pollution, water pollution, industrial air pollution, solid waste, hospital wastes, industrial hazardous wastes, noise pollution and people's participation in making Delhi a cleaner city. Directions have been issued by the National Capital Territory of Delhi for imposing restrictions on all commercial vehicles in Delhi in a time-bound programme beginning from April 1998. Based on the recommendations of a National Level Committee on Noise Pollution, directions have been issued to the State governments to check noise pollution from bursting of crackers. There is also a plan to introduce higher parking fees, to augment parking spaces in the city of Delhi and to introduce no traffic zones.

A detailed India Country Programme for phase out of Ozone Depicting Substances (ODS) was prepared in 1993 to ensure the phase out of ODS according to the national industrial development strategy, without undue burden to the consumers and the industry and for accessing the Protocol's financial mechanism in accordance with the requirements stipulated in the Montreal Protocol. At present, an exercise is underway in consultation with CII to update the country programme.

The main objectives of the Country Programme has been to minimize economic dislocation as a result of conversion to non-ODS technology, maximize indigenous production, give preference to one-time replacement, emphasize decentralized management and minimize obsolescence.

Environmental Impact Assessment

Impact assessment is a pointer to the environmental compatibility of the projects in terms of their location, suitability of technology, efficiency in resource utilization, recycling and so on. Impact assessment was introduced in India in 1978 and now covers projects such as:

(a) (i) river valley; (ii) thermal power; (iii) mining; (iv) industries; (v) atomic power; (vi) rail, road, highways, bridges; (vii) ports and harbors; (viii) airports; (ix) new towns and (x) communication projects;

(b) those which require the approval of the Public Investment Board/Planning Commission/Central Electricity Authority;

(c) those referred to the Ministry of Environment and Forests by other ministries;

(d) those which are sensitive and located in environmentally degraded areas; and

(e) public sector undertakings of the Centre where the project cost is more than Rs 500 million.

A notification issued in January 1994 makes Environmental Impact Assessment statutory for 29 categories of developmental projects under various sectors such as industrial, mining, irrigation, power, transport, tourism, communication, etc. The Environmental Impact Assessment (EIA) Notification was amended in 1997 to provide for public hearing as well as for empowering State governments for according environmental clearance in respect of certain Thermal Power Projects. As per the provisions of the EIA Notification 1994, mining projects are subject to environmental clearance.

However, it has been now decided to exempt prospecting through aerial survey and/ or test drilling in smaller areas from mandatory clearances. Applications complete in every material aspect are normally examined and decision conveyed to the applicants within 30 days in cases of site clearance and 120 days in cases of environmental clearance of projects. Once an application has been submitted by the project authority, the preliminary scrutiny of the project is done by the respective technical divisions and the overall appraisal of the projects is undertaken by specially constituted environmental appraisal committees of experts. In addition, special groups/committees and task forces are constituted as and when needed for expert inputs on major projects. After detailed scrutiny and assessment, the appraisal committee makes its recommendations for approval or rejection of the project. To ensure transparency, the position of forest and environmental clearance has been brought out on website: *httpc/www.nic.in/envfor* since February 1999.

Depending on the nature of the project, certain safeguards are recommended. For monitoring and timely implementation of safeguards suggested, six regional offices of the Ministry have been set up at Shillong, Bhubaneswar, Chandigarh, Bangalore, Lucknow and Bhopal.

OZONE CELL

India acceded to the Montreal Protocol, along with its London Amendment in 1992. To meet the country's commitment on ODS phase-out under the protocol and to disseminate information on ozone and ODS, the Ministry of Environment and Forests has established an Ozone Cell. A newsletter on ozone issues is being published by the Cell every two months and a number of workshops and seminars are also conducted to create awareness about ozone among industries. UNEP-IE Ozone Action Programme, Paris held its first south-Asia ODS officers Network Meeting in Delhi.

The Government provides custom/excise duty exemption for ODS phase-out projects and detailed guidelines/procedures have been finalized to grant duty exemption for new investments with non-ODS technologies. A policy to issue licenses for import of ODS has been implemented and the Reserve Bank of India, on the recommendation of the Ministry, has issued instructions to all commercial banks prohibiting finance or refinance of new investments with ODS technologies. One hundred seventy seven projects worth about US $ 49 million have been approved for India by the Multilateral Fund under the convention.

Trade in ODS with non-Parties has been banned. Imports and exports of ODS have been licensed. Exports of Chloro Fluro Carbons (CFCs) to developed countries has been stopped. Rules on Ozone Depleting Substances (ODS) phase-out were published on 9 May 1998 in the Gazette of India. The Policy to issue licenses for imports of ODS has been implemented.

The country thus has a strategy to phase out ODS by 2010 with minimum economic dislocation, and give preference to one-time replacement.

Short term Various ongoing programmes of energy efficiency and renewables other strategy highlights include:-

- To meet the freeze target of production and consumption of CFCs on 1.7.99.
- To meet the freeze target of production and consumption of halon in 2002.
- To reduce 50% consumption and production of CFC by 2005.
- To reduce 85% production and consumption of CTC by 2005.
- To complete phase out of production and consumption of CFCs, CTC, Halon by 2010.
- To complete phase out of Methyl chloroform by 2015.

Decision-Making: Major Groups Involvement

All the major groups are actively involved in the policy making process and decision making members from Business and industry, Scientific and Technical community, NGO's and women groups are members of the Empowered Steering Committees and various other advisory Committees. All groups have been affected. The insurance sector does not have a compensation package as yet for climate change.

Programmes and Projects

- Trade with non-Party has been banned.
- Fiscal measures have been taken by extending full exemption of duties (Customs and Excise) on goods required substitution of ODS and for establishment of new capacity with non-ODS technology.
- The Ozone Depleting Substances (Regulation and Control) Rules, 2000 have been notified.
- 225 ODS phase out projects have been approved by the Multilateral Fund to phase out 8600 ODP tones in India.

- Funding has also been approved for gradual phase out of production of CFCs.

The National Conservation Strategy and Policy Statement on Environment and Development, adopted by the Government of India in June 1992, lays down strategies and actions for integration of environmental considerations in the development activities of various sectors of the country, thus paving the way for achieving sustainable development. The action points pertaining to individual ministries/ departments ensure that they take action for reorienting their policies and programmes in conformity with the strategy. The activities for increase of forest covers and preservation of various kinds of eco systems eco systems are described below

Biosphere Reserves

Biosphere reserves are multi-purpose protected areas to preserve the genetic diversity in representative eco-systems. The major objectives of biosphere reserves are: (i) to conserve diversity and integrity of plants, animals and micro-organisms; (ii) to promote research on ecological conservation and other environmental aspects and (iii) to provide facilities for education, awareness and training. So far eleven biosphere reserves have been set up.

They are (i) Nilgiri; (ii) Nanda Devi; (iii) Nokrek; (iv) Great Nicobar; (v) Gulf of Mannar; (vi) Manas; (vii) Sunderbans; (viii) Similipal; (ix) Dibru Saikhowa; (x) Dehong Deband and (xi) Pachmarhi. Comprehensive guidelines for them emphasize formulation of eco-development and demonstration projects, development of data-base, conservation plans of key species, establishment of research stations and implementation of social welfare activities. Non-governmental organizations are being involved in the biosphere reserve programme for creation of public awareness.

Wetlands, Mangroves and Coral Reefs

India has a wealth of wetland eco-systems distributed in different geographical regions from the cold arid zone of Ladakh in the North to the wet humid climate of Imphal in the East, the warm arid zone of Rajasthan in the West to the tropical monsoon Central India and the wet and humid zone of Southern Peninsula. Most of the wetlands in India are directly or indirectly linked with major river systems such as Ganga, Brahmaputra, Narmada, Tapti, Godavari, Krishna, and Cauveri. A National Level Committee constituted to advise the Government on appropriate policies and measures to be taken for conservation and management of the wetlands, has so far identified 20 wetlands for conservation and management on priority basis. Steering Committees have been set up by the concerned State governments in which representatives of State government departments, universities and research institutions are included. Nodal research/academic institutions have been identified for each of the selected wetlands. Management Action Plans have been drawn up for most of the identified wetlands.

A directory on wetlands in India has been published which gives information on location, area and ecological categorization of wetlands in different parts of the country. India is a signatory to the Convention on Wetlands of international importance, especially as Waterfowl Habitat (Ramsar Convention) and six Indian Wetlands, viz., Keoladeo National Park, Bharatpur and Sambar (Rajasthan), Chilka (Orissa), Loktak (Manipur), Wullar (Jammu & Kashmir), and Harike (Punjab) have been designated under this Convention. Mangroves are salt-tolerant forest ecosystems found mainly in the tropical and sub-tropical inter-tidal regions of the world. They are reservoirs of a large number of plant and animal species associated together over a long evolutionary period and exhibiting remarkable capacity for salt tolerance. They stabilize the shoreline and act as a bulwark against encroachments by the sea.

India harbors some of the best mangroves in the world and these occur all along the Indian coastline in sheltered estuary, tidal creeks, backwaters, salt marshes and mud flats covering a total area of 4,827 sq km. Under the scheme on Conservation and Management of Mangroves, 15 mangrove areas have been identified for intensive conservation and management purposes: Northern Andaman and Nicobar, Sunderbans (West Bengal), Bhitarkanika (Orissa), Coringa, Godavari Delta and Krishna Estuary (Andhra Pradesh), Mahanadi Delta (Orissa), Pitchavaram and Point Calimer (Tamil Nadu), Goa, Gulf of Kutch (Gujarat), Coondapur (Karnataka), Achra/Ratnagiri (Maharashtra) and Vembanad (Kerala).

Management action plans for all the 15 mangrove areas have been sanctioned. Coral reefs are shallow-water tropical marine ecosystems, characterized by high biomass production and rich floral and faunal diversity. Four coral areas, Gulf of Mannar, Andaman and Nicobar Islands, Lakshadweep Islands and Gulf of Kuchch have been identified for conservation and management. State-level steering committees have been constituted for the formulation and implementation of management action plans. Such action plans have been sanctioned for Andaman and Nicobar and Gulf of Mannar coral reefs so far.

The objectives of the Convention on Biological Diversity (CBD) are: the conservation of Biological Diversity; the sustainable use of its component and the fair and equitable sharing of the benefits arising out of the utilization of genetic resources. Following the ratification of the Convention on Biological Diversity by India in 1994, several steps have been initiated both to meet the commitments under the Convention and to realize the opportunities offered by the CBD. A draft National Action Plan on Biodiversity has been finalized which seeks to consolidate the on-going efforts of conservation and sustainable use of biological diversity and to establish a policy and programme regime for the purpose. A legislation on bio-diversity is also under finalization. India regularly participates in important international conventions on Biological Diversity.

The problem of CO_2 emissions is a major concern to the Indian energy sector where coal accounts for over 60% of total energy resources used. In order to minimize $C0_2$ emissions, efforts are underway to improve efficiency levels in the generation and use of energy. In addition, renewable energy technologies and afforestation measures to increase the "carbon sink" function are being promoted. Coal India Limited (CIL), a holding company of seven coal producing companies, coordinates the implementation of sustainable development programmes in the Indian Coal Sector. There is a special focus on ensuring conservation of coal sources during exploitation and use, and conserving energy in the production and transportation of coal.

As sugar is a major industry in India, the potential for power generation through bagasse based co-generation is estimated at 3,500 MW. The programme will set up capacity for the generation of 300 MW based on bagasse within the next three years. Demonstration projects on biomass based power generation units are being carried out at the block level. The projects use locally available biomass including agro-waste for power generation. Several programmes for the recovery of energy from urban, municipal, and industrial wastes, and alternate energy for transportation, and tapping of ocean energy have also been initiated.

Status

Preliminary assessment of the Indian coast and its vulnerability to sea level rise reveal that in terms of the total number of people at risk is 7.1 million persons on a nationwide basis. The simulation studies for wheat showed that an increase in temperature to be caused a severe decrease in yield. Impact in general is being evaluated and assessed regularly.

Current Indian gross Carbon dioxide emissions on a per capita basis is merely 1/6th of the world average.

Fresh land availability is sparse as already 4% of the geographic area is under nature reserves / forest land. The Deptt. of Space in India has mapped the entire country on a 1:1million scale to identify various land use categories. This study has shown that wasteland area is around 53.3 mn hectares. The remote sensing forestry data can be used for GHG sink development and impact assessment of land use change.

The consumption of CFC has been reduced to 60% of the base-level consumption. Use of halon has been reduced to 20% of the consumption of the base-year.

Preliminary studies on the impact of a rise in sea level of 1 mm on the Indian coastline indicate that 0.41% of India's coastal area will be inundated. Some studies suggest that as the greenhouse effect gains strength, the cyclones will become more frequent and more destructive making island archipelagos such as Lakshadweep highly vulnerable. The danger of frequent storms which generally originate in the Bay of Bengal is, however, higher in the Andamans and Nicobar than Lakshadsweep. Also, as sea level rises, the fresh water aquifers of the islands will be subjected to saline intrusion.

The total installed power capacity in India is 80,000 MW with a per capita consumption of about 300 KWHours/year. Thermal and conventional hydro power contribute about 96% of the total installed capacity. India has a total renewable energy potential of about 126,000 MW (wind 20,000 MW, micro-hydro 10,000 MW, biomass/bioenergy 17,000 MW, ocean thermal power 50,000 MW, tidal power 9,000 MW, and sea wave power 20,000 MW). In addition, India receives a total solar insulation of the equivalent of 5 times 101 KWHours/year. Besides the potential energy that can be derived from these sources, there is a drive to achieve the target of installing 12 million family type biogas plants and 120 million improved cookstoves that will support the cause of energy conservation. Several major renewable energy programmes are being undertaken in these areas across the nation. Over 2.2 million biogas plants and 22 million improved wood stoves have been installed in rural and remote areas resulting in the saving of the equivalent of 21 million tonnes of fuel wood per annum. Moreover, the biogas plants are producing about 30 million tonnes of enriched organic manure per annum.

Power plants of 15-100 MW capacity based on biomass are being established and bio-fuels used to generate electricity and for thermal application. So far over 1,500 biomass gasifiers with aggregate capacity equivalent to 16 KW have been set up. Considering the resources available in the country in the form of agricultural residue, agro-industrial residue, and wasteland for energy plantations, the total exploitable energy potential in the country has been estimated at about 17,000 MW.

In order to meet the basic lighting requirements in rural and semi-urban areas, about 32,000 solar street lights, 30,000 domestic lights, and 37,000 solar lanterns have been made available. Solar photovoltaic systems are also being used for a variety of other applications in rural areas. A total of about 1.25 solar photovoltaic systems with a total capacity of 14 MW have been installed under the programme. The projects are expected to add 202 MW to installed capacity.

Challenges

The metros and other big cities of the country are the worst affected chiefly because of lopsided urban planning and incorrect location of industrial units. However, a strategy for corrective action has been initiated.

The more vulnerable weaker sections of society find it more difficult to cope.

Lack of new and additional financial resources as per the provisions of UNFCC

Lack of transfer of ESTs to India by the developed world countries on grant / preferential/ confessional terms

India's efforts at protecting the ozone layer are guided by the need to integrate environmental protection with development while formulating policy and implementing programmes. The objectives of the ozone depleting substances (ODS) phase-out need to be achieved with minimum economic dislocation and minimum obsolescence costs. Indigenous production of products and substitutes require encouragement, and technological choices need to be carefully made. The special requirements of small and medium enterprises will be addressed. This many be achieved by a mix of instruments that include information dissemination, fiscal measures, regulations, etc. The implementation of ODS phase-out programmes will however be contingent upon the availability of assistance, including technology, from the Multilateral Fund.

Capacity-building, Education, Training and Awareness-raising Education, Awareness and Information

Priority is accorded by the Ministry of Environment and Forests to promote environmental education, create environmental awareness among various age-groups and to disseminate information through Environmental Information System (ENVIS) network to all concerned. A major initiative to include environment education as a separate and compulsory subject in the educational curricula has been taken by the Ministry at all levels of formal education, i.e., secondary, senior secondary and tertiary levels. A discussion paper prepared on strengthening environment education was presented by the Minister for Environment and Forests at the State Education Ministers' Conference held from 22 to 24 October 1998. The paper was adopted by the Conference. The Chief Ministers were urged to introduce environment education in the school curricula from the 1999-2000 academic session. Maharashtra is the first State to introduce the subject in the school curriculum.

Paryavaran Vahinis (environment launch-vehicle) are proposed to be constituted in 194 selected districts all over the country which have a high incidence of pollution and density of tribal and forest population. The Vahinis also play a watch-dog role by reporting instances of environmental pollution, deforestation and poaching. They function under the charge of District Collectors, with the active cooperation of the State/Union Territory governments. This scheme is entirely financed by the Ministry of Environment and Forests. Seven Centres of Excellence have been set up by the Ministry to strengthen awareness, research and training in priority areas of Environmental Science and Management. These are: Centre for Ecological Sciences, Bangalore; Centre for Mining Environment, Dhanbad; Centre for Environmental Education, Ahmedabad; CPR Environmental Education Centre, Chennai; Salim Ali Centre for Ornithology and Natural History, Coimbatore; the Centre for Environmental Management of Degraded Ecosystems, Delhi, and the Tropical Botanical Garden and Research Institute, Thiruvananthapuram, Kerala.

The National Museum of Natural History (NMNH) set up in New Delhi in 1978, is concerned with the promotion of non-formal education in the area of environment and conservation. Besides permanent exhibit galleries on various aspects of environment, the museum also conducts temporary exhibitions and a large number of educational programmes and activities for school children, college youth and the general public. Three Regional Museums of Natural History have been established at Mysore, Bhopal and Bhubaneswar.

The Indian Council for Forestry Research and Education is the focal point for forestry education and extension development in the country. The Indira Gandhi National Forest Academy, Dehra Dun, imparts in-service professional training to Indian Forest Service (IFS) professionals. State forest service colleges provide training to the officers of the State Forest Service (SFS). The Indian Plywood Industries Research and Training Institute, Bangalore, organizes short-term courses in the area of wood science. The Indian Institute of Forest Management, Bhopal, also provides training in forest management and allied subjects to persons from the Indian Forest Service, forest development corporations, and forest-related industries to develop forestry programmes. The Wildlife Institute of India, Dehradun, provides in-service training to forest officers, wildlife ecologists and other professionals for conservation and management of the wildlife resources of the country.

Ozone Layer Protection

Information dissemination package for school teachers and NGOs prepared by Centre for Environment Education was launched on 16th September, 1998 and distributed in four workshops organized in Calcutta, Delhi, Pune and Chennai in November-December, 1998. This kit has been developed in consultation with the UNEP-DTIE office in Paris.

Painting Competitions are being organized by the Ozone Cell. An Indian entry won the prize in the international competition organized by UNEP in 1999. A National Painting Competition has been organized on the occasion of Sixty International Ozone Day, 16th September, 2000.

A car sticker is being brought out for distribution every year on the International Ozone Day.

Ozone-friendly equipment and products are being exhibited during 16-18th September every year. A similar exhibition is being held on the occasion of the International Ozone Day, 2000.

An information kit on Ozone Science and Ozone layer has been prepared by Centre for Environment Education, Ahmedabad. Chapters on pollution and environmental issues are part of the curriculum.

Several training programmes are being organized on a regular basis. India is a member of WMO and is participating in various international research programmes on systematic observation atmosphere and ocean through IMD/DST New Delhi.

Information

The India Meteorological Department (IMD) was established in 1875. It is the National Meteorological Service and the principal government agency in all matters relating to meteorology, seismology and allied subjects. The Department has units all over the country engaged in collecting meteorological and seismological data besides providing various meteorological services. Its main objective is to provide meteorological information for weather sensitive activities like aviation, shipping, agriculture, irrigation, off-shore oil exploration and industries. The Department also issues warnings against severe weather phenomena like cyclones, dust-storms, heavy rains, cold and heat waves that cause destruction of life and property. Besides, it also provides climatological information, records earthquakes and promotes research in meteorology. The Department maintains an extensive network of modern observatories and communication links all over the country. Observations received through high power radars and weather satellites are extensively used these days for analysis and prediction of weather.

INSAT Meteorological Data Processing System is being upgraded to handle reception, processing of data from INSAT 2E Satellite. This satellite is similar to INSAT 2B but has got additional capability of providing imagery in water-vapor band and higher resolution imagery in visible, near IR and short wave IR bands, using charge coupled devices. Current satellites provide imagery in Visible and IR bands. A satellite based wide area network (WAN) using Very Small Aperture Terminals (VSATs) is being established for speedy dissemination of forecast products to various user agencies, particularly to AAS units. The plan is being implemented by installing VSATs in 127 locations out of which 47 have been installed. Scientific campaigns such as GAME-India (Global Energy Water Cycle Asian Monsoon Experiment), I-STEP (Indian Solar Terrestrial Energy Programme), OCEANSAT (IRS P4), Indo-US Agreement on Satellite Data Utilization, and INDOEX are being carried out.

The Ozone measurement is being carried out at 5 stations by Indian Meteorological Department. These stations are at Srinagar, New Delhi, Varanasi, Pune and Kodaikenal. The total Ozone data and Umkehr data (vertical profile of Ozone) are being regularly sent in WMO format to the World Ozone Data Centre, Canada and are being regularly published by the Centre. India developed ozonesonde in 1963. Since 1970, vertical Ozone distributions are being measured at 5 stations at New Delhi, Pune, Thiruvananthapuram, Dakshni Gangotri and Maitri.

Scientific data and information on the protection of atmosphere and ozone Layer are made available to potential users at national level via internet and other modes of information dissemination.

Information on Ozone Science and non-ODS technology has been provided by UNEP DTIE Ozone Action Programme through its clearing housing mechanism and they send information directly to users. There are several programmes for awareness and information dissemination and websites exist for access to information.

Research and Technologies

There are no carbon dioxide monitoring stations. These are expensive and are required for early detection concerning changes in the atmosphere. Also in their presence it is easy to discern other competing / modulating effects in the atmosphere.

The Government of India is making several efforts for developing new environmental friendly technologies for example, Dobson and Brewer Spectrophotometers are used to monitor total ozone measurement.

The Government of India is disappointed by the non implementation of commitments made by the developed countries on providing new and additional financial resources as well as technology for promoting ecologically sustainable development. In order to implement measures to protect the atmosphere, India need the following new technologies:

- IGCC and other advanced clean coal technologies;
- state of art power generation technologies;
- Solar photovoltaic (advanced systems);
- Mass rapid transport system for urban transport; New non-ODS technology for refrigeration, aerosol, foam fire extinguisher and solvent sector; and
- Technology for manufacturing ODS substitutes.

Producers of CFCs, and the Government of India jointly supported research programmes to develop technology to produce HFC 134a at the Indian Institute of Chemical Technology, Hyderabad. The pilot plant has been set up. Commercial production of HFC-134a is envisaged.

Both Government and voluntary organizations are involved in climate change research in India with the former supporting a wide variety of projects in the area of global change research. The India Meteorological Department (IMD) monitors climate to detect change, predict climate change, determine the effects of climate change, and contribute to global observational efforts. Since 1983, IMD has maintained a meteorological observatory at the Indian permanent station in Antarctica. The Department has actively participated in various international and National observational efforts. A National Climate Centre (NCC) has been established at IMD's research wing in Pune to undertakes climate research, applications, data collection and management, and impact awareness studies. Data collected from the extensive network of observations are archived at the Department's National Data Centre in Pune. The Centre holds over 60 million records in its archives. About 2.5 million records are currently being added to the archives.

The Government of India through the Department of Science and Technology (DST) has undertaken several projects to promote technologies that will reduce pollution of the atmosphere. Technologies that use compressed natural gas (CNG) as fuel in internal combustion engines have been developed for generating electricity in remote hilly areas. To help promote waste reduction, the production of fuel pellets from waste has been successfully demonstrated to industrial users in Mumbai. DST supports R&D efforts to promote environmental conservation conducted by several autonomous institutions and service organizations. Considerable research work has been completed on climate modeling, air pollution, and atmospheric ozone at the Institute of Tropical Meteorology, Pune, which has climate change as one of its major thrust areas.

Financing

Activities concerning protection of Atmosphere cut across several government departments and the main source is public funding. Despite lack of external funding India has made big strides in protecting the atmosphere.

Climate change/ protection of atmosphere considerations have been in built into the five year plan process in the country and an adequate thrust has been accorded to it by GOI

Ozone regulations and Market forces are driving the industries to invest money in non-ODS technologies.

Cooperation International Co-operation

The Ministry of Environment and Forests functions as a nodal agency for United Nations Environment Programme (UNEP), South Asia Co-operation Environment Programme (SACEP), and International Centre for Integrated Mountain and Development (ICIMOD), International Union for Conservation of Nature and Natural Resources (IUCN) and various international agencies, regional bodies and multilateral institutions. India is signatory to the following important international treaties/ agreements in the field of environment:

(i) International Convention for the regulation of Whaling;

(ii) International Plant Protection Convention;

(iii) The Antarctic Treaty;

(iv) Convention on Wetlands of international importance;

(v) Convention on International trade in Endangered Species of Wild Flora and Fauna;

(vi) Protocol of 1978 relating to the international convention for the prevention of pollution from ships;

(vii) Vienna Convention for the protection of the Ozone Layer;

(viii) Convention on Migratory Species;

(ix) Basel Convention on Trans-boundary movement of hazardous substances;

(x) Framework Convention on Climate Change;

(xi) Convention on conservation of bio-diversity;

(xii) Montreal Protocol on the substances that deplete the ozone layer; and

(xiii) International Convention for Combating Desertification.

The Ministry and its agencies cooperate with various countries such as Sweden, Netherlands, Norway, Denmark, Australia, U.K., U.S.A., Canada, Japan, FRG, etc., on bilateral basis and with several UN and other multilateral agencies such as UNDP, World Bank, Asian Development Bank, OECF (Japan) and ODA (U.K.) for various environmental and forestry projects.

India is a party to GEF and has following UNDP/GEF projects

i) GHG stabilization in India

ii) Enabling activities for preparation of an initial national communication

The Article 3 of the Vienna Convention for protection of the Ozone Layer provide for conducting research and scientific assessment..

India is a party to UNFCCC. It was among the first fifty countries to bring it into force w.e.f March 21,1994

India ratified the Montreal Protocol along with its London Amendment in 1992. India is yet to ratify the Kyoto Protocol. The Montreal Protocol under its article 10 has the provision to provide financial assistance to the developing countries. A multilateral Fund has been set up in 1992. Developed countries replenish the Fund by their annual contribution. This may continue till 2010.

The Government of India is disappointed at the non implementation of two important / seminal commitments by the developed countries at UNCED 1992 relating to New and additional financial resources Transfer of EST's. It is a matter of regret there is no progress on transfer of EST's to developing countries.

- Technology for development of ODS substitute chemical is need to be transferred in fair and favorable terms.
- The policies adopted by the Executive Committee of the Montreal Protocol do not address the special needs of small and medium enterprises. It is, therefore, special funding wind with differentiated cost effectiveness for all sectors in SMEs is need to be established.

BIODIVERSITY

Decision-Making: Legislation and Regulations

Conservation and sustainable use of biological resources based on local knowledge systems and practices is ingrained in Indian ethos and way of life. Formal policies and programmes for conservation and sustainable use of biodiversity resources date back several decades. The concept of environmental protection is enshrined in the Indian Constitution in Article 48(a) and 51(g). Major central acts relevant to biodiversity are: the Wildlife (Protection) Act, 1972; the Forest (Conservation) Act, 1980; and the Environment (Protection) Act, 1986. The various central acts are supported by a number of state laws and statutes concerning forests and other natural resources.

Decision-Making: Strategies, Policies and Plans

Following the ratification of the Biodiversity Convention by India, several steps have been initiated to meet the commitments under the Convention and to bring the legislative, administrative, and policy regime regarding biological diversity in tune with the Articles of the Convention. A National Action Plan on Biological Diversity is being finalized which will consolidate the ongoing conservation and sustainable use efforts including capacity building and biosafety measures. In addition, the following activities are being undertaken: biosafety protocol; biodiversity information network; capacity building in taxonomy; consultation with the State Governments; traditional knowledge and benefit sharing; and legislation.

Policies and strategies directly relevant to biodiversity include: the National Forest Policy as amended in 1988; the National Conservation Strategy, and Policy Statement for Environment and Sustainable Development; the National Agricultural Policy; the National Land Use Policy; the National Fisheries Policy (under preparation); the National Biodiversity Policy (under preparation); the National Wildlife Action Plan; and the Environmental Action Plan.

Programmes and Projects

A programme called eco-development for in situ conservation of biological diversity involving local communities has been initiated in recent years through World Bank assistance. The concept of eco-development integrates the ecological and economic parameters for the sustained conservation of ecosystems by involving local communities with the maintenance of earmarked regions surrounding protected areas. To conserve representative ecosystems, a Biosphere Reserve Programme is being implemented. Eight biodiversity rich areas of the country have been designated as Biosphere Reserves applying the United Nations Educational, Scientific and Cultural Organization's Man and the Biosphere (UNESCO MAB) criteria. India is one of the 12 mega biodiversity centres in the world, representing two of the major realms and three of the basic biomes of the world. The country is divided into 10 biogeographic regions: Trans-Himalayan, Himalayan, Indian Desert, Semi-Arid, Western Ghats, Deccan Peninsula, Gangetic Plains, North-East India, Islands, and Coasts.

Programmes have also been launched for scientific management and wise use of fragile ecosystems. Specific programmes for management and conservation of wetlands, mangroves, and coral reef systems are being implemented. National and state level committees oversee and guide these programmes to ensure strong policy and strategic support.

Status

Approximately 5.3% of the total geographical area of country has been earmarked for extensive in situ conservation of habitats and eco-systems through a protection area network of 80 National Parks and 44 Wildlife Sanctuaries. This network has played a significant role in restoring viable populations of large mammals such as the tiger, lion, rhinoceros, crocodiles, elephants, etc.

Attention has been paid to ex situ conservation measures as they complement and are important to situ conservation. According to a recent survey, the Central and State Governments together run and manage 33 Botanical Gardens. In addition, universities have their own Botanical Gardens. There are 275 centres of ex situ wildlife preservation in the form of zoos, deer parks, safari parks, aquaria, etc. The Government of India has set up a Central Zoo Authority to oversee, monitor, and coordinate the management and the development of zoos in the country.

Challenges

Research and Technologies

Surveys of the floral and faunal resources in the country are carried out by the Botanical Survey of India established in 1890 and the Zoological Survey of India established in

1916. The National Institute of Oceanography and several other specialized institutions and universities further strengthen the taxonomic data base. The diversity of the country's biological resources is yet to be fully surveyed. Approximately 65% of the total geographical area has been surveyed to date. Based on this, over 47,000 species of plants and 81,000 species of animals have been recorded. The vascular flora which form the conspicuous vegetation cover comprises about 15,000 species. Several thousands of them are endemic to India and they have so far not been reported from anywhere else in the world. This list is being constantly upgraded, especially in respect of lower plants and invertebrate animals. The biological diversity of the country is so rich that it may play a very important and crucial role in the future survival of mankind if it is conserved and used with the utmost care. Today, two hot spots in biological diversity have been identified in the country, namely, the Eastern Himalayan region and the Western Ghats. A comprehensive status report covering the various facets of biodiversity conservation in India is under preparation.

The collection and preservation of genetic resources is accomplished through the National Bureau of Plant Genetic Resources for Wildlife for crop plants; the National Bureau of Animal Genetic Resources for domesticated animals; and the National Bureau of Fish Genetic Resources for economically valuable fish species. These Bureaus are assigned the task of collecting germplasm and supplying these on request to Indian and foreign agencies for research purposes.

Cooperation

The Convention on Biological Diversity was signed and ratified by India in February, 1994. The Convention on International Trade in Endangered Species of Wild Fauna and Flora was signed in 1976.

India believes that National action regarding conservation and sustainable use of biodiversity and equitable sharing of benefits arising from the use of genetic resources demands appropriate actions on the part of international community. Some key issues in this regard are as follows:

1. Development of a suitable enabling environment by the other parties, particularly the developed countries, to ensure benefits to countries of origin. These benefits should not only include measures like royalty payment or monetary compensation, but also location of research and technologies in the countries of origin in accordance with the provisions of the Convention.
2. Development of a credible internationally regime for recognizing the intellectual and physical property rights of local communities. Development of such a regime may take time, pending which all patent applications should be required to disclose: a) the source and origin of the genetic material used; b) knowledge and practices about the use of the genetic resources by the local communities and identification of such communities; and c) a declaration that laws, practices, and guidelines for the use of such material and knowledge systems in the country of origin have been followed.
3. Capacities of biodiversity rich countries should be built to enable them to bio-prospect and develop products from genetic resources.
4. Introduction of transgenics, alien species should be only permitted with appropriate safeguards.

This information was provided by the Government of India to the 5th Session of the United Nations Commission on Sustainable Development. Last Update: 1 April 1997.

DESERTIFICATION AND DROUGHT

Decision-Making: Strategies, Policies and Plans

In the National Conservation Strategy particular attention has been paid to arid and semi-arid areas. The strategy, inter alia, includes classification, zoning, and apportionment of land for designated uses; enactment of laws for appropriate land uses to protect the soil from erosion; pollution and degradation measures for runoff and wind erosion; development of suitable agro-silvipastoral techniques; measures for water conservation, recycling and optimal conjunctive use of surface and ground water; and encouragement for and improvement in traditional methods of rain water harvesting.

Programmes and Projects

The Desert Development Programme (DDP) was initiated in 1977-78. It covers both the hot desert regions of Gujarat, Rajasthan, and Haryana and the cold desert areas in Jammu, Kashmir, and Himachal Pradesh. It is functional in 131 blocks of 21 districts in 5 States covering an area of about 0.362 million km^2 and a population of 15 million. The objectives of the programme include controlling the process of desertification, mitigating the effects of drought, restoring the ecological balance, and raising the productivity of land, water, livestock, and human resources. At least 75% of the allocation is earmarked for activities which would contribute towards combating the process of desertification. The programme is implemented with 100% central assistance. The Programme Evaluation Organization of the Planning Commission has the task of evaluating this programme in order to assess its impact on the control of desertification, and on improvements in productivity and income for the people living in these areas. From 1990 to 1993, Rs. 1,485 million has been spent under the scheme, developing an area of 90,412 ha.

The Drought Prone Area Programme (DPAP) was launched in 1973 in arid and semi-arid areas with poor natural resource endowments. The objective is to promote more productive dryland agriculture by better soil and moisture conservation, more scientific use of water resources, afforestation, and livestock development through development of fodder and pasture resource, and in the long run to restore the ecological balance. The DPAP covers 615 blocks of 91 districts in 13 States. This is a Centrally sponsored scheme where the allocations are shared between the Centre and States on a 50:50 basis. Preparation of development plans on a watershed basis, participation of people in planning and implementation of the programme, and developing effective liaison between research agencies and implementing agencies are some of the priority areas of the programme which is being implemented under the Eighth Five Year Plan with renewed thrust. From 1990 to 1993, Rs. 3,066.9 million has been spent by the scheme, developing an area of 571,633 ha.

In order to integrate and intensify the activities aimed at combating desertification, a comprehensive plan for control of desertification under the National Forestry Action Programme has been proposed. The plan would evaluate the present status of deserts in the country, assess the implementation of ongoing programmes for development of deserts and desert prone areas, formulate broad policy guidelines and action plans for implementation aimed at control of desertification, develop strategies involving people in desert control through various means, and include appropriate measures related to research and training in desert control.

The basic objective of the integrated Wastelands Development Project is to facilitate pilot projects using an integrated approach to wasteland development by initiating area-specific projects taking into account land capabilities, site condition and local needs, and ultimately aiming to promote optimal land use for both ecological and socioeconomic needs. The different types of problem lands for which projects are prepared include saline/alkaline lands, arid/sandy areas, ravine areas, and Aravallis. The activities covert soil and water conservation, afforestation, silvi-pasture development, grazing management, etc.

The main objective of the Afforestation Project for the Aravalli (Rajasthan) is to check desertification and restore ecological status by re-afforestation and also to increase the production of fuel wood, fodder, timber, and non-wood forest products to meet local needs. The project started in April 1992 and the project period is 5 years.

Rehabilitation of common lands in Aravallis (Haryana) is being implemented in the four southern districts of Haryana, that is Bhiwani, Mahendragarh, Gurgaon, and Faridabad since 1990. The project outlay is Rs. 480 million and covers environmental protection, restoration of green cover in the semi-arid Aravalli Hills, and improvement in the living conditions of the local people through meeting their biomass needs.

Status

About 10% of the 329 million ha of land area in India is arid. This zone is located in the western region. Rajasthan accounts for 61% of the arid area, and a further 20% is located in the adjoining State of Gujarat. Cold deserts located in the High Himalayas of the North West account for the rest. Semi-arid areas account for 30.56% of the area and are located in 127 districts of 10 States. There is a well defined desert region consisting of the great desert and little desert. The great desert extends from Rann of Kutch beyond Luni river northwards. The little desert is located between Jodhpur and Jaisalmer and the two are divided by a zone of sterile rocky land cut up by limestone ridges.

There is evidence that the arid area once had dense forest cover. Large scale migrations through the North Western Himalayan passes resulted in clearance of natural vegetation for settled agriculture. As arid areas are located on migration routes, the delicate balance of water and nutrient recycling was lost with the indiscriminate spread of agriculture which started around three thousand years ago. The river "Saraswati" of Indian mythology vanished altogether while other rivers merged into the sand dunes.

The semi-arid regions abutting the arid zone on the North and North East have a better water regime as a number of perennial rivers fed by Himalayan snow traverse this area.

A well knit irrigation system makes the area the most productive part of the country. In the East and South, however, agriculture is mainly rain-fed particularly in the plateau region. Periodic cycles of drought due to monsoon failure is a common feature.

Research and Technologies

Research activities pertaining to various aspects of arid zones are being conducted at the Central Arid Zone Research Institute (CAZRI). The majority of these activities are oriented towards agriculture and soil conservation. CAZRI, located in Jodhpur, was established in 1988 under the auspices of Indian Council of Forestry Research and Education (ICFRE) with the prime objective to complete research into sand-dune stabilization, afforestation of the arid saline land of Rann of Kutch, Aravalli Hills, and the Indira Gandhi Nahar Project (IGNP) command area, appropriate land use systems, silviculture of important shrubs and trees with emphasis on selection and tree improvement, vegetative propagation, etc. Important studies that have been conducted include the identification of species most suitable for restricting the movement of sand-dunes and checking the advance of the desert, the influence of moisture conservation practices in the establishment of plantations in arid and semi-arid areas, investigation of the influence of farm yard manure and nitrogen and potassium fertilizers on the establishment and growth of Prosopis cineraria and Tecomella undulata, irrigation management in forestry plantations in the IGNP command area of the Indian Desert, and the Combined Production System (Agri-silvi-pastoral) in arid regions. India has built up some degree of expertise in matters relevant to desertification. These include long range weather forecasting, remote sensing, research in arid zone agriculture, forestry and pastures, and dry land farming.

Cooperation

The International Convention to Combat Desertification in Countries Experiencing Drought and/or Desertification Particularly in Africa was signed by India on December 26, 1996.

India has been participating regularly in the Inter-governmental Negotiating Committee to Combat Desertification (INCD) process. In collaboration with the Interim Secretariat of the United Nations Convention to Combat Desertification, the Government of India hosted a Regional Conference on the Implementation of the UN Convention to Combat Desertification and Drought in Asia from August 21-23, 1996 at New Delhi. At the meeting, the countries resolved to initiate consultations among themselves to identify specific programmes for regional cooperation.

ENERGY

Decision-Making: Coordinating Bodies

The names of the Government Ministries/offices having overall responsibility for making policy decisions concerning energy issues in general and energy-related aspects of atmosphere and transportation include the Ministries of:

- Power, Petroleum and Natural Gas, Non-Conventional Energy Sources, Environment, and Surface Transport;
- Department of Coal;
- Department of Atomic Energy; and
- Central and States Pollution Control Boards

While each Ministry or Department frames the policy for the sector it administers, the Ministry of Power is responsible for overall coordination for energy-related matters. At the official level, the highest coordinating body is the Committee of Secretaries where each Ministry is represented by its Secretary. There are various other Committees which also help in inter-Ministerial coordination like Working Groups which are constituted for the Five Year Plans and the Coal Linkage Committee which coordinates the addition to thermal installed capacity for power generation with that of coal demand/supply.

The Ministry of Environment separately assesses the environmental impact of new power stations through various inter-ministerial Committees. Inputs are also provided by the Planning Commission, which is the highest federal body for planning purposes including for allocation of resources. The Energy Policy Unit in the Planning Commission which studies and analyses the inter-sectoral issues on energy. Effort is made at all levels to integrate the views of Business and Industry, the Science and Technology community, local authorities and NGO'S.

The Constitution of India clearly demarcates the areas which would be under the authority of the Federal Government and that of State Governments. The Federal and State Governments are jointly responsible for some activities though in case of any inconsistency, the Federal Law prevails. Electricity is a Concurrent subject at entry 38 in list III of the Seventh Schedule of the Constitution of India. The Ministry of Power is primarily responsible for the development of electrical energy in the country. The Ministry is concerned with perspective planning, policy formulation, processing of projects for investment decision, monitoring of the implementation of power projects, training and man-power development and the administration and enactment of legislation in regard to thermal and hydro power generation, transmission and distribution. On energy related issues, states have their state electricity boards (SEB's) and departments.

The Government of India made amendments to the Indian Constitution; the Seventy Third and Seventy Fourth amendment in the early 1980s identified specific areas of work which could be taken up by local governments. All the states have a nodal agency for implementation of renewable energy programmes. The specific priorities within the state for various programmes are also decided.

Decision-Making: Legislation and Regulations

Regarding Power Supply, the following Acts are in operation:

(i) The Indian Electricity Act, 1950

(ii) Electricity Supply Act, 1948

(iii) Electricity Regulatory Commission Act, 1998

The following Acts are applicable for environment protection relating to the atmosphere:

(i) The Environment Protection Act, 1980.

(ii) The Air (Prevention and Control of Pollution) Act, 1981.

M/o Road Transport & Highways deals with Indian Motor Vehicle Act, 1988 and Central Motor Vehicle Rules, 1989. The Motor Vehicle Act (Amendment), 2000 legislated the use of environment-friendly fuel like Compressed Natural Gas (CNG) and Liquefied Petroleum Gas (LPG) as auto-fuels. In addition, Bharat Stage-I norms, which are akin to Euro-I norms, have been introduced all over the country w.e.f. 1.4.2000. Further, Bharat Stage-II norms, which envisage a Sulphur content of 0.05% as against higher quantities under Bharat Stage-I, have been introduced in NCR of Delhi, and these are being introduced in the other three metropolitan cities namely Mumbai, Calcutta and Chennai in a phased manner

In addition to legislation, there are various incentives and award schemes to promote sustainable development:

- National Energy Conservation award for various types of industries;
- Reward schemes for meritorious performance which include efficient operation of Thermal Power Station (TPS);
- Incentive award for improved Station Heat Rate of TPS;
- Schemes for installation of energy saving lamps, computerized load management, installation of Time-of-Day energy meters, rectification of agricultural pump sets, etc.;
- Incentives offered for installation of electrical gadgets deriving energy from renewables; and
- Schemes for System Improvement and Transmission and distribution loss reduction.

The major policy initiatives taken to encourage private/ foreign direct investment to tap energy from renewable energy sources, include provision of fiscal and financial incentives under a wide range of programmes being implemented by the Ministry of Non Conventional Energy Sources, and simplification of procedures for private investment, including foreign direct investment in renewable energy projects.

There is also a package of incentives for renewable energy projects. These include:

- Concessional/nil customs duty on import of projects, equipment and components related to renewable energy;
- 100% depreciation allowed in the first year of investment for the installation of renewable energy projects (except for small hydro projects);
- liberalized foreign investment approval regime to facilitate foreign investment and transfer of technology through joint ventures. Proposal for up to 100% foreign equity participation in a joint venture qualify for automatic approval; and
- Policy announcements by state governments/ SEB's for evacuation of power generated from renewable energy projects with facilities for wheeling, banking, third party sale and purchase of power by SEB's at remunerative prices.

The fiscal incentives provided for this purpose include 100 per cent depreciation in the first year of the installation of the project, exemption from excise duty and sales tax and concessional customs duty on the import of material, component and equipment used in renewable energy projects. In addition, the Government provides financial incentives, such as interest subsidy and capital subsidy from the Ministry and soft loans from Indian Renewable Energy Development Agency (IREDA). Fourteen states have so far announced such policies in respect of various renewable energy sources.

6

SOCIAL ASPECTS OF SUSTAINABLE DEVELOPMENT IN INDIA

POVERTY

Decision-Making: Strategies, Policies and Plans

Poverty eradication remains the overriding priority for India. The challenge is to find a development path that is not only sustainable but is also socially just and culturally acceptable. India has set a target for the eradication of absolute poverty by the year 2002.

India has a three pronged strategy for poverty eradication: economic growth and overall development; human development with emphasis on health, education and minimum needs, including protection of human rights and raising the social status of the weak and the poor; and directly targeted programmes for poverty alleviation through employment generation, training, and building up the poor's asset endowment. Removal of poverty has always been one of the main objectives of India's Five Year Plans. The Eighth Five Year Plan of India (1992-97) recognized human development as the ultimate goal of the development process. Employment generation, population control, illiteracy, education, health, provision of drinking water, and adequate food are listed as priorities towards the achievement of this goal.

Many of the programmes and activities outlined in the Programme of Action adopted at the World Summit for Social Development held in Copenhagen in 1995 are already in place in India, particularly the policies geared towards eradication of poverty, generation of employment, etc. India has taken a range of measures to implement the Programme of Action at the National level. India has established a National Committee for Social Development in the Planning Commission, and the State Governments are also being encouraged to establish similar committees. While the primary responsibility for implementation of the Programme of Action rests with the States, the National effort will need to be strengthened and supplemented by the efforts of the international community. It is, therefore, necessary for the international community to dedicate itself to the task of fulfilling the commitments undertaken at Copenhagen.

The pro-poor components of the poverty plan are fully integrated with the overall development plan of the country and the pro-poor perspective is fully harmonized with the strategy of market-oriented open-economy industrialization and the requirements of structural adjustments. Economic growth enables expansion of productive employment and also generation of resources which are vital to support any form of intervention for eradication of poverty. Since 1991, India has undertaken trade reforms, financial sector reforms, and removal of controls and bottlenecks. These reforms were introduced with the objective of improving efficiency and productivity, in order to further accelerate growth by improving competitiveness in international markets. The ultimate objective of such reforms is to ensure expeditious eradication of poverty. Adequate precaution was taken to protect the poorer sections of society against short term effects of these changes. This has been done through increasing the resources for programmes for the poor in the National Plan and sharpening the focus of such programmes on the poor.

The Conference of Chief Ministers on Basic Minimum Services held at New Delhi during 4-5 July, 1996, recommended the adoption of the following objectives with an all out effort for their attainment by the year 2000: 100%

provision of safe drinking water in rural and urban areas; 100% coverage of primary health service facilities in rural and urban area; universal primary education; provision of Public Housing Assistance to all shelterless poor families; extension of the Mid-day Meal Programme in primary schools to all rural blocks, and urban slums and disadvantaged sections; linkage provisions to all unconnected villages and habitations; and streamlining the Public Distribution System targeted to families below the poverty line.

The Conference recommended that all centrally sponsored schemes relating to the above seven Basic Minimum Services should be continued and the States' annual entitlement should be increased by 15-20% every year. It also recommended that the funds allocated for these Basic Minimum Services in the States' and the Central Plan should not be diverted. The Budget for 1996-97 provides an additional Rs. 24.66 billion with a view to increase the availability of funds for State level social programmes for safe drinking water, primary education, primary health, housing, mid-day meals for primary school children, rural roads, and strengthening public distribution system.

Programmes and Projects

The Government of India has adopted various schemes and programmes for accelerating the rate of economic growth, eradication of rural poverty through wage employment and self-employment, redistribution of land and security of land tenure, enhanced Minimum Needs Programme, protection of minorities, availability of opportunity for socioeconomic uplift, and infrastructure development to help the urban poor.

The Common Minimum Programme (CMP) announced by the Government in June, 1996 has shown strong commitment to the development of social sectors for achieving distributive justice. The Government has also accorded high priority to poverty alleviation programmes. The Central Plan allocations for social sectors and poverty alleviation programmes show the highest increase in 1996-97 over 1995-96. The weaker sections of society have been given importance in special programmes of poverty alleviation and employment as well as in several other programmes. The Public Distribution System has been recently streamlined in order to target the poorer sections of the population. The poverty alleviation programme is one of the main thrust areas of the Common Minimum Programme. A strategic attack on poverty is an important element of the development policy pursued by the Government. A two-pronged attack on rural and urban poverty has been launched through wage employment and credit linked self-employment schemes.

Some of the new initiatives in the area of poverty eradication and social sector include: a) National Social Assistance Programme (NSAP) which covers National Old Age Pension Scheme (NOAPS), National Family Benefit Scheme (NFBS), and National Maternity Benefit Scheme (NMBS); b) Nutritional Support Primary Education (NSPE), a mid-day meal scheme for school children in all Government, local body, and private aided schools (classes I to V); c) Indira Mahila Yojana with three components: convergence of inter-sectoral activities, income generating activities, and sustained process of awareness generation/education; d) Pension Scheme for Provident Fund Subscribers (PSPFS); e) Social Security for Construction Workers; f) Rural Group Life Insurance Scheme which incorporates a 50% subsidized policy available to one member of a rural poor family below the poverty line; f) Prime Minister's Integrated Urban Poverty Eradication Programme; g) Pulse Polio Immunization Programme; h) Revamped Rural Employment Programme encompassing the Employment Assurance Scheme (EAS), Jawahar Rozgar Yojana (JRY), Indira Awaas Yojana (IAY), and the Million Wells Scheme (MWS); and i) Revamped Integrated Rural Development Programme.

Status

The estimated proportion of population below the poverty line is sensitive to the estimation procedure used. However, according to a variety of estimates, the decline in the proportion of population below the poverty line was between 2.9 and 8.7% in the six year period ending 1993-94. The appropriate methodology for estimation of poverty is also currently under review by the Planning Commission.

Cooperation

India believes that poverty anywhere is a threat to prosperity everywhere and that concerted international action is essential to ensure global prosperity and better standards of life for all. Based on this belief, India has actively played a positive, constructive role in the deliberations of the UN, its specialized agencies, and various intergovernmental mechanisms. The Rio Declaration on environment and development adopted at the United Nations Conference on Environment and Development (UNCED) in 1992 states, inter alia, that "eradicating poverty and reducing disparities in living standards in different parts of the world are essential to achieve sustainable development and meet the needs of the majority of people." Agenda 21, the blueprint for sustainable development, emphasizes that the actions of individual governments in combating poverty require the support of the international community as the struggle against poverty is a shared responsibility of all countries. A favourable international economic environment, combined with financial and technical assistance, favourable terms of trade, debt relief, access to markets, and transfer of environmentally sound technologies, will help pave the way for poverty eradication and sustainable development. Poverty has so many causes that no single solution will solve the problem in all countries.

DEMOGRAPHICS

Decision-Making: Coordinating Bodies

In keeping with the move towards integration of social sector services in India, a coordination mechanism has been established at the Central level, and States have been advised to set up similar mechanisms for State/Union Territories, district, and block levels.

Decision-Making: Strategies, Policies and Plans

Population stabilization is an essential prerequisite for sustainable development. In India, the National Family

Planning Programme was launched in 1952 with the objective of "reducing birth rate to the extent necessary to stabilize the population at a level consistent with requirement of the National economy". The technological advances and improved quality and coverage of health care has resulted in a rapid fall of the mortality rate from 27 in 1951 to 9.8 in 1991. In contrast, the reduction in birth rate has been less steep, declining from 40 in 1951 to 29.5 in 1991. As a result, the annual exponential population growth has been over 2% in the last three decades. During the Eighth Five Year Plan period (1992-97), the fall in birth rate has been steeper than that of the death rate; consequently the annual growth rate was around 1.9% during 1991-95. The rate of decline in population growth is likely to be accelerated during the Ninth Five Year Plan period (1997-2002).

India participated in and is a party to the Programme of Action (POA) of the International Conference on Population and Development (ICPD), held at Cairo in 1994. In keeping with the global vision of population programmes contained in the Programme of Action, a paradigm shift has taken place in the National Family Welfare Programme. From April 1996, this Programme is being implemented on the basis of the "Target Free Approach (TFA)". This approach is based on replacement of the system of setting contraceptive targets from the top by a system of decentralized participatory planning at the Primary Health Centre (PHC) level.

In keeping with the policy shift in the population programme, the Ninth Plan will contain a large project, the "Reproductive and Child Health Project" (RCH), which contains two components. The first will be a National component to continue the Child Survival and Safe Motherhood (CSSM) project. The CSSM, which has been extended to cover all districts of the country, aims to provide universal immunization to children against six vaccine-preventable diseases, namely polio, pertussis, tuberculosis, measles, tetanus, and diphtheria, as well as immunization against tetanus for pregnant women. Other activities are control of diarrhoea—a major cause of infant and child mortality, anaemia in children and women, blindness in children, and acute respiratory infections in children. Under the safe-motherhood component, enhancement of the percentage of delivery by trained personnel, training of traditional birth attendants, provision of emergency obstetric care, etc. are being provided.

Decision-Making: Major Groups Involvement

The number of government assisted NGOs participating in population programmes during 1995-96 was about 900. During the current year, 1996-97, about 650 NGOs have been given assistance up to January, 1997. This excludes private-sector-for-profit-providers (individual doctors and facilities), who also provide a wide range of reproductive and child health services.

Programmes and Projects

Major public health interventions were undertaken in the years 1995 to 1997. In keeping with the goal of eradication of polio by the year 2000 AD, the first round of Pulse Polio Immunization (PPI) was completed for children in the 0-3 year age group in December, 1995 and January, 1996. On the first (December 9, 1995) National Immunization Day, 87 million children were immunized and in the second (January 20, 1996), 93 million children were immunized against polio. The exercise has been repeated on December 7, 1996 and the coverage in 490 of the 510 districts which have reported so far is 115 million children.

Similarly, a check-up for primary school children was carried out over the period July/October, 1996. Coverage of children enrolled in schools was about 85%. The objectives of this check-up was to detect common health problems among school children, refer those needing treatment, and build health awareness in the community.

Both these initiatives attracted tremendous community response. Based on the experience of the PPI, the Department of Family Welfare has plans to use the Pulse Polio Immunization Campaign (PPIC) posts on a regular basis for reproductive and child health care including immunization. The check-up of school children has since been mandated by law to become a regular annual feature, particularly for detection of incipient disabilities. The screening leads to referral and treatment as necessary.

Status

Decentralized participatory planning stresses quality of care, and assessment and provision of services on the basis of client needs. Voluntary and informed choice, which has always been the underpinning of the population programme, is stressed. Service providers/ managers at the level of the Primary Health Centre are expected to draw up a health care and family welfare plan at their level after carrying out assessment of community needs and in consultation with the community and community leaders. An integrated system of monitoring has been devised and transmission of information will be done using the country-wide governmental information network known as the NIC-NET. While the reporting system is still to be put into place, it is expected that it will relieve service providers of excessive record keeping and free their valuable time for providing services to and making contact with the community. The mobility of service providers is being enhanced, to facilitate greater and more frequent contact with the community. As many as 132,285 sub-centres at the peripheral level have a female para-medical [known as the Auxiliary Nurse Midwife (ANM)], and the medical doctors at the Primary Health Centre (PHC) level include a fair number of women. The women's perspective is expected to be taken care of by this empowerment of service providers and managers.

Capacity-building, Education, Training and Awareness-raising

Information Education and Communication (IEC) is carried out through a variety of modes of dissemination, including interpersonal communication, electronic and print media, animation groups and peer groups, population education through schools and colleges, adult literacy campaigns, youth clubs, etc. Required software for IEC is designed and supplied. Efforts are being made to build capacity at the State level to design and create software more appropriate to local needs. IEC has been diversified from being contraception oriented to encompassing issues like the status of women, child survival, age at marriage, etc. Correspondingly, messages pertaining to fertility are also being carried in the IEC projects of other Departments such as Education and Rural Development. Training of all service providers at the block level will be integrated from April 1, 1997 to facilitate appreciation of the interlinkages between various aspects of development like literacy, poverty alleviation, and environment with population.

Research and Technologies

Extensive research has been undertaken in India on the reproductive health of the population. The National Family Health Survey (NFHS), conducted in 1992-1993, covered 89,777 married women in the 13-49 year age group in 88,562 households in 24 States and the Union Territory of Delhi. The Survey has yielded a vast amount of data, which is being used for policy and programme purposes.

Financing

It is estimated that in the Ninth Plan period, external assistance for the population programme will be at a level of about Rs. 30 billion. Agencies like the United Nations Population Fund (UNFPA) and the United Nations Children's Fund (UNICEF) have and are continuing to provide valuable technical, monetary, and material assistance. Bilateral agencies are providing financial support for the population programme. The PPI initiative has attracted widespread international support.

HEALTH

Decision-Making: Strategies, Policies and Plans

The strategy of Indian Health Planning is two pronged; first, to build up a primary health care infrastructure, and second, tackling specific diseases. The primary health care infrastructure, consisting of Sub-centres, Primary Health Centres (PHCs), and Community Health Centres, has been built around a population of 30,000 per unit. This mechanism provides for a sustained and continuous outreach for all health and family welfare programmes in the country. The disease specific strategy consists of programmes aimed at prevention and control of specific diseases. These programmes are targeted for specific regions depending upon circumstances/spread of the disease.

The focus of the Eighth Five Year Plan for the Health Sector has been to improve access to health care for the under-served and under privileged segments of the population. This is being achieved through: a) consolidation and implementation of the primary, secondary, and tertiary health care infrastructure for optimal performance, and building up appropriate referral services with emphasis on primary care; and b) effective implementation of National programmes for combating major public health problems.

Until poor sanitation, contaminated water supply, and lack of adequate facilities for solid and liquid waste management both in urban and rural areas are corrected, it may not be possible to completely prevent periodic outbreaks of infectious diseases. Nevertheless, if outbreaks are detected early enough, it will be possible to control the epidemic and reduce morbidity and fatality rates. The strategy during the Ninth Plan will be to strengthen health surveillance, and early alert and rapid response mechanisms at district and sub-district levels. This would necessitate the provision of epidemiological expertise and diagnostic laboratory services as an essential component of existing health care system.

It is neither possible nor feasible to initiate and support vertical programmes for control of every non-communicable disease. During the Ninth Plan period, integrated noncommunicable disease control programmes will be implemented using the experience gained from pilot projects, such as the diabetes control programme launched during the Eighth Plan.

Urban migration over the last decade has resulted in the rapid growth of urban slums. In some cities, the health status of urban slum dwellers is worse than that of the rural population. During the Ninth Plan period, steps will be initiated to develop a well structured organization of urban primary health care to ensure basic Health and Family Welfare dwellings. Appropriate referral linkages between primary, secondary, and tertiary care facilities in defined geographic areas will be established to promote optimal use of all available facilities. Increasing involvement of the Nagar Palikas in the implementation of health, water supply, and sanitation programmes is expected to improve the health status of the urban population, especially slum dwellers and those living below the poverty line.

Programmes and Projects

Some of the National Health Programmes are: National Malaria Eradication Programme; National Tuberculosis Control Programme; National AIDS Control Programme; and National Blindness Control Programme. The "Health for all" strategy is being re-oriented towards Health for the Under Privileged. In view of the importance given to medical and health care in the economic reforms, the Central Plan outlay for programmes of the Department of Health has been increased.

The National Health Programmes aimed at prevention, control and eradication of communicable and non-communicable diseases have been accepted by the Government for implementation. Efforts have been made to ensure that the ongoing reforms do not lead to any adverse effect on the provision of essential care to meet the health needs of disadvantaged segments of the population. Some of the measures include allocation of funds under the Social Safety Net Scheme to improve Maternal and Child Health (MCH) infrastructure beginning in a phased manner with 90 poorly performing districts.

Under the Basic Minimum Services scheme, the Government is committed to providing credible primary health care at the 5000 population level. In addition to the Centrally Sponsored Schemes, funds will be devolved to States for meeting requirements under the Basic Minimum Services, including health and family welfare. The strengthening of rural health infrastructure has been undertaken over the years by the Department of Family Welfare through the provision of buildings, equipment, drugs, vaccines, and training at all personnel levels.

For AIDS, a National Control Programme has been established with blood safety measures and sexually transmitted disease (STD) control through the National AIDS Control Organization. During the last four years of implementation of the programme, 154 Zonal Blood Testing Centres have been established all over the country to provide HIV testing facilities. In total, 199 blood banks were modernized during 1995-96. One hundred and twenty eight medical officers, 747 blood bank technicians, and 37 drug inspectors have undergone training under the Programme. The "One World, One Hope" theme that was adopted for the World AIDS Day on December 1, 1996 reflects the coming together of various groups to prevent the spread of HIV.

Status

Communicable diseases continue to be a major cause of morbidity and mortality in India. In addition to the existing bacterial, viral, and parasitic infections, there are newer additions such as HIV infection and re-emergence of some infections such as kala azar, so that the disease burden due to communicable diseases continues to be very high. There are National Programmes for control of vectors, but performance in many of these has been sub-optimal, an important factor being the lack of key personnel such as lab technicians and multipurpose workers. Many of these programmes were initiated at a time when primary health care infrastructure was not fully operational and, hence, had their own vertical infrastructure. In the Ninth Five Year Plan period, a major effort will be initiated for horizontal integration of these programmes at the district and sub-district levels within the existing framework of primary health care infrastructure.

Challenges

Changing lifestyles, longevity, and dietary habits have resulted in increased prevalence and earlier age of onset for diabetes, and cerebro- and cardio-vascular diseases over the last decade and a concomitant rise in the disease burden and disability adjusted life years (DALY) due to non-communicable diseases. The overall cancer incidence in the country is low. Even though the two common cancers of the oropharynx and uterine cervix are easy to diagnose and treat, the available data indicate that the majority of cases are detected at a late stage when palliative rather than curative treatment remains the only possible therapeutic modality. Thus, there is a need to improve the facilities for early detection of cancers so that effective treatment could be provided.

Capacity-building, Education, Training and Awareness-raising

There are over 0.65 million Indian Systems of Medicine (ISM) practitioners in the country. They work in remote rural as well as urban slum areas and could play an important role in enhancing health care outreach. There is a need to improve pre-service training and provide periodic updating after graduation so that there is improvement in the quality of service and greater participation in meeting the health care needs of the population. It is important to increase the efficiency of the health system through all categories of health manpower during the Ninth Plan period. The recommendations contained in the National Education Policy on Health Sciences, as approved by the Central Council of Health and Family Welfare in 1993, will be implemented to ensure growth and development of the appropriate mix of health care manpower. Optimal use of human resources for health will be made through: creation of a functional, reliable health management information system, and training and deployment of health managers with requisite professional competence; multi-professional education to promote team work; skills upgrading for all categories of health personnel as part of structured continuing education; increasing accountability of responsiveness to people's health needs by assigning an appropriate role to the Panchayati Raj institutions; and making use of available local and community resources.

Financing

To augment the resources for health care, user charges have been introduced for medical/diagnostic services in certain hospitals in some of the States except for the poor. This will help to provide better quality services, besides facilitating public funding of basic health facilities. With rising incomes, the demand for health care is increasing.

EDUCATION

Decision-Making: Legislation and Regulations

The goal of universal elementary education is enshrined in the Constitution of India. It will be India's goal to ensure that all school-going children in the 6-14 year age group are enrolled by the year 2000. India's system of elementary education is the second largest in the world with 151 million children enrolled in the 6-14 year age group in 1994-95. This covered about 91% of the children in this age group. The important requirement of supportive infrastructure is included under the "Operation Blackboard" scheme started in 1987-88.

Decision-Making: Strategies, Policies and Plans

The National Policy of Education (NPE), 1986 aims at illiteracy eradication in the 15-35 age group by the year 2000.

Eradication of illiteracy (EOI), has been accorded a high priority in the Eighth Five Year Plan (1992-97) and is a major priority area. It is also one of the components of the Minimum Needs Programme (MNP). The National Literacy Mission (NLM) was launched in May, 1988 for achieving universal literacy in the 15-35 age group. The target is to cover 100 million adult illiterate persons during the Eighth Plan period. The emphasis is on sustainability of literacy skills and the achievement of goals of remediation, continuation, and application of skills to actual living conditions. This programme also concentrates on education for weaker sections of society like the members of Scheduled Castes and Tribes, and women. The NLM will achieve the coverage of 100 million adults by the year 1998-99 with special emphasis on the spread of literacy among women and in the States with high incidence of illiteracy.

The Total Literacy Campaign has become the principal strategy of the NLM in the eradication of illiteracy throughout the country. The target for achieving total literacy is now 2005 AD. As on December 1996, about 417 districts have been covered either fully or partially under this Campaign. Similarly, 178 districts have been covered either fully or partially under the Post Literacy Campaign. Under all the schemes of NLM, 57.96 million persons, out of an enrolment of 96.80 million, have so far been made literate according to NLM norms. The new scheme of Continuing Education for neo-literates is now under implementation (approved in December 1995).

In the Ninth Five Year Plan, making the nation fully literate by the year 2005 will be a committed goal. This carries out the directions of the 1992 update of the NPE. Around 6% of the GDP will be earmarked for the education sector by the year 2000 and 50% of that will be spent on primary education. Further, substantial funds will be earmarked for technical and vocational training, in order to generate more employable and self-employed youths.

Programmes and Projects

The main problem faced in the implementation of the programme of elementary education is the high drop-out rates. Efforts have been made to counter this by providing free elementary education, with a scheme of free textbooks and uniforms. A scheme for mid-day meals (Nutritional Support to Primary Education) has been recently launched. The scheme will be implemented in all the States to ensure regular attendance and retention in primary and middle schools. In every initiative to promote the spread of education, the girl child will be a special focus of attention. The District Primary Education Programme (DPEP) which became operational in 1994-95, attempts to take a holistic view of primary education development and seeks to implement the strategy of universality of elementary education, through district planning and desegregated target setting.

Status

Human resource development, one of the most important needs of India, is receiving the desired attention. The important components of human resource development include education, training, awareness raising, dissemination of information, and decision making.

EDUCATION IS VIEWED AS A FUNDAMENTAL HUMAN RIGHT.

Capacity-building, Education, Training and Awareness-raising

Environmental education forms an essential ingredient in the education process of the country. The National Policy on Education provides for including environment, among other factors, as an integral part of the curricula at all stages of education. The National Council of Educational Research & Training (NCERT), New Delhi has developed syllabi and curricula on environmental education both for Primary and Secondary School levels.

A comprehensive document "Environmental Education in the School Curriculum" has been published which lists the approach and concepts covered in different subjects at different stages of 12-year schooling. During 1996-97, a National Resource Centre in Environmental Education was also been established to promote better awareness, understanding, and sharing of experiences and materials in environmental education. An effort is being made to develop a data bank of various institutions, activities, and materials in environmental education to facilitate better interaction and dissemination among the different agencies and the school system.

This country-wide coverage of science exhibitions being organized at the District, State, and National levels has helped in spreading environmental awareness and motivating children to think about the control measures for environmental protection. The NCERT has also developed audio-video programmes on various related themes of environmental education and sustainable development for school children and teachers.

Region-specific training modules for District Institutes of Educational Training rich in environmental concepts have been developed. Environmental education is made an essential component of training programmes for teachers and teacher educators/trainers. The State Boards of Education have started follow up action for development of curricula with environment education as an important element. NGOs are also being encouraged by financial assistance from the government to produce experimental and innovative work in the field of environment education.

Financing

The share of both Central and State Governments, including local bodies, in financing educational institutions continues to be quite high. It accounted for 92.9% of the total income of educational institutions in 1990-91, whereas the share from fees, endowments, and other sources declined sharply. While the Central Government plays an important role for overall policy directions in education and funding of centrally sponsored schemes, the State Governments provide most of the funding for the education system. Private initiatives in education may be encouraged to supplement public resources. Further, resources must be mobilized by revising fees and other user charges especially for higher levels of education.

HUMAN SETTLEMENTS

Decision-Making: Legislation and Regulations

Land remains the most critical constraint in the development of the housing sector, particularly in the larger cities. Legislative provisions like the urban land ceiling, rental laws, and planning codes are among the major constraints and conservative land use norms have restricted the supply of land into the market. Apart from these constraints, there are substantial vacant land holdings in the possession of government departments, educational institutions, religious and charitable trusts, and corporations. Bringing these holdings into the land market would help augment land supply.

Decision-Making: Strategies, Policies and Plans

As one of the original signatories to the Vancouver Action Plan, 1976, India has introduced approaches in its human settlements programmes that seek to effectively provide access for people, especially vulnerable groups, to adequate and affordable shelter in human settlements that encompass the shelter unit and basic physical, economic, and social services, including access to livelihood programmes. India initiated the process through formulation of the National Housing Policy (NHP) with the long term goals to reduce the number of homeless, to improve the housing conditions of the inadequately housed, and to provide a minimum level of basic services and amenities to all. The foundation that has been strongly established over the last 20 years, enables the activities to gather momentum and take the directions that are considered necessary to implement the Habitat II (Istanbul, 1996) National Plan for Action (NPA).

Major priority issues are identified in the NPA. The objective is to create the enabling environment in which participants outside the government system can become more active in the delivery of housing solutions and provision of services, so that the outreach is extended to all segments of the market, especially vulnerable groups. The NPA, a consensus effort of all the key actors, has two critical objectives, namely, giving people access to adequate and affordable shelter and social infrastructure and services, and developing sustainable urban and rural settlements in an urbanizing world.

The NPA specifically encompasses the following major activities: creation of an enabling environment; development of all types of housing and related services; eradication of poverty and strengthening the activities in the informal sector; providing access for women, children, and other vulnerable groups to housing and basic services; monitoring and evaluation systems; and State shelter policies and action plans. All the key actors are committed to the implementation of the NPA and the Global Plan of Action, to which the NPA is closely linked. The Government of India reaffirms its commitment to realize the rights set out in relevant international instruments and documents relating to education, food, shelter, employment, health, and information, particularly in order to assist people living in poverty. The strategy of the Ninth Five Year Plan (1997-2002) is to provide housing for all by the terminal year of the Plan.

Programmes and Projects

In spite of rapid and widespread urbanization, India still has a large rural population, 629 million, living in 580,706 villages. The average population of an Indian town is 60,297 and that of an Indian village 1,083. The attractiveness of rural development programmes has been a contributory factor for villages with over 10,000 population preferring to remain in the rural category. These include, in particular, the Integrated Rural Development Programme (IRDP), Rural

Labour and Employment Generation Programme (RLEGP), Jawahar Rozgar Yojna (JRY), and Indira Awaas Yojna (IAY), which have improved housing conditions, income opportunities, and accelerated economic growth. Rural areas have contributed to the sustainability of urbanization by providing inputs for urban industry, trade and services, a large market for urban products, a source of competitively priced labour, and household savings to the financial system. The strengthening of the rural-urban continuum is high on the habitat agenda for India.

Status

The current state of human settlements in India presents a mixed scenario. There have been significant improvements in the coverage of the population's basic human settlement amenities and in the quality of the habitats. There has been a visible improvement in housing structure and quality and more market-sourced materials are being used in both urban and rural areas. Higher levels of affordability have been achieved. At the same time, housing costs are rising, floor area per capita is falling, and a growing number of people are being pushed out of the formal housing market. The impact of the situation is reflected in the proliferation of urban slums.

INSTITUTIONAL ASPECTS OF SUSTAINABLE DEVELOPMENT IN INDIA INTEGRATED DECISION-MAKING

Decision-Making: Coordinating Bodies

In India, the National Environmental Council is the key sustainable development coordination mechanism. The Council, chaired by the Prime Minister, is the highest policy making body on environmental issues. The Council consists of senior representatives of Central Ministries, Chief Ministers of States, representatives of Non-Governmental groups, distinguished scientists, and academics.

Decision-Making: Legislation and Regulations

India's development objectives as reflected in the planning process have consistently emphasized the promotion of policies and programmes for economic growth and social welfare. The alleviation of poverty and the development of the country's economic and social infrastructure have been emphasized in the country's successive Five Year Plans. Investment resources were targeted to ensure the realization of these concerns. Environmental issues which have been an integral part of Indian thought and social processes are reflected in the Constitution of the Republic of India adopted in 1950.

Decision-Making: Strategies, Policies and Plans

The Directive Principles of State Policy enunciate principles which, though not enforceable by any Court, are nevertheless fundamental in the governance of the country. It is the duty of the State to apply these principles in legislation. The commitment of the State to protect environment and safeguard forests and wildlife is reflected by specific provisions in the Directive Principles of State Policy. Further, the Constitution states that it shall be the fundamental duty of every citizen to protect and improve the natural environment, including forests, lakes, rivers, and wildlife, and to have compassion for living creatures. By a 1976 Constitutional amendment, the subject of forests and wildlife was brought under the Concurrent List in the Seventh Schedule, thereby enabling Parliament and the Central Government to legislate on these subjects. The roots of the growing trend towards popular participation in the conservation and natural resource development programmes lie in these Constitutional provisions.

The National Conservation Strategy and Policy Statement on Environment and Development, adopted in June, 1992, provide the basis for the integration and internalization of environmental considerations in the policies and programmes of different sectors. It also emphasizes sustainable lifestyles, and the proper management and conservation of resources. The policy statement announced by the Government in 1992 on Abatement of Pollution reiterates the Government's commitment to arrest deterioration of the environment. The statement reflects a shift in focus from problems to implementation of measures incorporating both short-term and long-term considerations. The statement recognizes that pollution particularly affects the poor. The complexities are considerable given the number of industries, organizations, and government bodies involved. To achieve the objectives, maximum use is made of a mix of instruments including legislation and regulation, fiscal incentives, voluntary agreements, educational programmes, and information campaigns.

Enabling people to identify their own strengths and weaknesses and equipping them with the necessary skills and capabilities is an important step in their empowerment. Voluntary action helps this process. Traditionally, voluntary organizations have played an important role in India. The Eighth Five Year Plan had identified people's initiative and participation as a key element in the process of development. It had also recognized that the role of the Government should be to facilitate and strengthen the process of involvement of major groups by creating the right types of institutional infrastructure.

The Ninth Five Year Plan (1997-2002) is being launched in the 50th year of India's Independence. The objectives of the Ninth Plan arising from the Common Minimum Programme of the Government are as follows: priority to agriculture and rural development with a view to generating adequate productive employment and eradication of poverty; accelerating the growth rate of the economy with stable prices; ensuring food and nutritional security for all, particularly the vulnerable sections of society; providing the basic minimum services of safe drinking water, primary health care facilities, universal primary education, shelter, and connectivity to all in a time-bound manner; containing the growth rate of population; ensuring environmental sustainability of the development process through social mobilization and participation of people at all levels; empowerment of women and socially disadvantaged groups such as Scheduled Castes, Scheduled Tribes, and other backward classes and minorities as agents of socioeconomic change and development; promoting and developing participatory institutions like Panchayati Raj institutions, cooperatives, and self-help groups; and strengthening efforts to build self-reliance.

Decision-Making: Major Groups Involvement

There are over 10,000 NGOs in India ranging from National agencies to local groups, from research organizations to mass-based field organizations. Many of these are engaged in promoting eco-development, waste management, forest conservation, preservation of genetic diversity, and eco-friendly technologies in industry and agriculture. Voluntary agencies have developed a variety of innovative approaches that could help secure the involvement of local communities, particularly the poorer sections, in various developmental activities. Voluntary organizations have largely been responsible for ensuring the better delivery of rural services that include drinking water facilities, sanitation, road development programmes, etc. The costs of providing basic services have consequently been reduced due to the successful mobilization of local resources at low cost for implementation of development programmes. The Council for Advancement of People's Action and Rural Technology (CAPART) is the agency for financing and assisting voluntary action in the area of rural development.

The adoption of the 72nd and 73rd Constitutional Amendments in 1992 by Parliament is a landmark event in the lives of Indian women, as they are assured one-third of the total seats in all elected offices in local bodies both in rural and urban areas. As a result of this, women have been brought to the centre-stage in the nation's efforts to strengthen democratic institutions at the grassroots levels and to enter into public life through 2,30,000 local bodies all over the country.

Programmes and Projects

The need to integrate the environment and development decision making process has been recognized as contributing to economically efficient, socially equitable, and responsible environmental management. More extensive use of analytical tools, such as environmental impact assessment (EIA) and environmental health impact assessment (EHIA) on strategic policies and development programmes which have an adverse effect on environment or on human health, environmental risk assessment (ERA) of industrial units, and environmental audit (EA) to increase efficiency in the use of energy and resources and reduce wastes can contribute to policy integration by making decision makers aware of the environmental consequences of their actions.

A very far-reaching notification by the Ministry of Environment and Forests gazetted in 1994 makes it obligatory for almost all development activities, small and large, to conduct an environmental impact assessment study which has to be evaluated and assessed by an impact assessment agency (Ministry of Environment and Forests) who may consult a Committee of Experts, if deemed necessary. The assessment shall be completed within a period of 90 days and the decision on the approval conveyed within 30 days after completion of public hearings when required. No developmental activity can be taken up unless the conditions stipulated under the respective environmental and forestry clearance have been complied with. Environmental protection cannot be isolated from the general issues of development and must be viewed as an integral part of development efforts. Accordingly, the concept of sustainable development must include the fostering of economic growth, the meeting of basic domestic needs (including health, nutrition, education, housing, etc.), and the eradication of poverty so as to provide to all a life of dignity in a clean, safe, and healthy environment. Stress needs to be placed equally on the "development" and "sustainable" dimensions of the concept of sustainable development. The integration of environmental concepts into policies and programmes concerning economic development should be carried out without introducing a new form of conditionality in aid or development financing. Having achieved substantial progress in terms of software and hardware requirements, the emphasis from now on should be on enforcement and performance evaluation of assets created. A monitoring mechanism will, therefore, need to be reoriented in the case of major projects and programmes to achieve this objective.

Status

India is of the view that there is no conflict between environment and development. Various efforts have been made to integrate environmental concerns into the decision making process. Environmental standards and environmental management plans are important measures taken to protect the environment. The same applies to environment audits which are being made mandatory for major industries.

An important element of sustainability pertains to the protection of the environment and preservation of the natural resource base of the nation. Rapidly growing population; urbanization; changing agricultural, industrial, and water resource management; and increasing use of pesticides and fossil fuels have all resulted in perceptible deterioration in the quality and sustainability of the environment. It should be realized that environmental protection does not only involve a prevention of pollution and of natural resource degradation, but has to be integrated with the overall development process and the well-being of people.

National Decision-Making Structure

1. National Sustainable Development Coordination Body: YES
2. National Sustainable Development Policy: YES

3. National Agenda 21/other strategy for SD: IN PROCESS
4. Local/Regional Agenda(s) 21: NO
5. Environmental Impact Assessment Law: YES
6. Major Groups involved in Sustainable Development Decision-Making: YES

National Instruments and Programmes

1. Sustainable. Dev. or Environmental education incorporated into school curricula: YES
2. Sustainable Development Indicators Program: YES
3. Ecolabel Regulations: YES
4. Recycle/Reuse Programs: YES
5. Green Accounting Program: YES - Municipalities
6. Access to Internet: YES
7. Access to World Wide Web: YES
8. A national World Wide Web Site for Sustainable Dev. or State of the Environment: YES

Policies, Programmes, and Legislation

1. Combatting poverty: YES
2. Changing consumption and production patterns: YES
3. Atmosphere: YES
4. Land Use Planning: YES
5. Forest and Deforestation: YES
6. Desertification and Drought: YES
7. Sustainable Mountain Development: YES
8. Sustainable Agriculture: YES
9. Biological Diversity: YES
10. Biotechnology: YES
11. Oceans and Coastal Areas: YES
12. Freshwater Management: YES
13. Toxic Chemicals: YES

14. Hazardous Wastes: YES
15. Solid Wastes: YES
16. Radioactive Wastes: YES
17. Energy: YES
18. Transport: YES
19. Sustainable Tourism: YES

Capacity-building, Education, Training and Awareness-raising

In India, the need for community participation in development activities has been fully appreciated and recognized. It is realized that developmental activities undertaken with the active participation of major groups have a greater chance of success and can also be more cost effective. In the area of education, health, family planning, land improvement, efficient land use, minor irrigation, watershed management, recovery of wastelands, afforestation, animal husbandry, dairy and sericulture, considerable progress has been achieved by creating institutions for people and encouraging community participation.

The Ministry of Environment and Forests has instituted "Paryavaran Vahinis" from 1992-93 with the basic objective of creating environmental awareness through people's participation. In addition, 3,000 eco-clubs have been set up in schools with the Ministry's assistance.

MAJOR GROUPS WOMEN

Decision-Making: Legislation and Regulations

India believes that gender issues cannot be solely left to market forces and remain the responsibility of both National governments and the international community. Gender equity is a necessary pre-condition for fulfillment of the goals agreed to at the Beijing Conference. India's Constitution and legal framework uphold the dignity and status of women and seek to create an environment where empowerment is facilitated. A National Commission for Women has been established and the National Human Rights Commission has the mandate to examine human rights issues involving women. A Commissioner for Women's Rights has also been appointed. Special cells for preventing crime against women are also available. The Equal Remuneration Act stipulates payment of equal remuneration to men and women workers for work of equal value. The Act also prohibits any gender discrimination in recruitment and service conditions. The 73rd and 74th Constitutional Amendment Bills, adopted in 1992 by Parliament to strengthen the Panchayati Raj System (the system of self-governance at the local level both in rural and urban areas), provide that a third of all elected offices in local bodies (rural and urban) are reserved for women.

Decision-Making: Strategies, Policies and Plans

The concept of women's development in the initial Five Year Plans was mainly welfare oriented. In the Fifth Plan (1974-79), however, there was a shift in the approach towards women from 'welfare' to 'development'. The new approach aimed at an integration of welfare with developmental services. The Sixth Plan (1980-85), adopted a multi-disciplinary approach with a three-pronged thrust on health, education, and employment. In the Seventh Plan (1985-90), the developmental programmes for women continued with the major objectives of raising their economic and social status and to integrate them better with mainstream National development. The Eighth Plan (1992-97), promises to ensure that the benefits of development from different sectors specifically benefit women, and that these programmes be implemented to complement the overall programme. Women must be enabled to function as equal partners and participants in the development process. This approach marks a shift from 'development' to 'empowerment'.

The Ninth Plan (1997-2002), with the empowerment of women as one of its major objectives, will create an enabling environment with requisite policies and programmes, legislative support, exclusive institutional mechanisms at various levels, and adequate financial and human resources to achieve this objective. An integrated approach will be adopted towards empowering women. This underscores the harmonization of various efforts on different fronts, that is, social, economic, legal, and political. Further, a special strategy of earmarking funds as 'women's component' will also be adopted with a close vigil to ensure an adequate share of resources and benefits for women from all developmental sectors both in the Central and State Sectors. To this effect, the Ninth Plan recommends prompt adoption of the formulated National Policy for Empowering Women along with a well defined Gender Development Index to monitor the impact of its implementation in raising the status of women. During the Ninth Plan, a Media Policy will be framed to project a positive image of the girl child and women. A strict ban on the depiction of demeaning, degrading, negative, and conventional stereotypical images of women and violence against women will be enforced through legislation, regulatory mechanisms, and media policies.

Two new initiatives were launched for women during the Eighth Plan. The first, Mahila Samridhi Yojana (MSY) launched in 1993, attempts to promote the habit of savings. The second scheme, the National Credit Fund for women called Rashtriya Mahila Kosh (RMK), was set up in March 1993 to meet the credit needs of poor women and particularly those in the unorganized sector, who would otherwise have difficult access to formal institutional credit instruments. Within a short span of three years, the RMK had sanctioned extended credit limits of Rs. 260 million to 129 NGOs for its further lending to individuals. This will benefit over 136,000 women. Out of this amount, Rs. 162.2 million has already been disbursed

Some of the other activities undertaken during the Eighth Plan include, adopting a National Plan of Action for children and the girl child; establishing the National Creche Fund for child care services; adopting a National Nutrition

Policy (1993); and integrating the Child Development Service (ICDS). ICDS is a major programme for the provision of nutrition needs for mother and child. This scheme covers the welfare of children below the age of six and expectant and lactating mothers.

India is committed to increase investment on education to 6% of GDP, with the major focus on women and the girl child. The Government is committed to universal mother and child care programmes to reach out to every corner of the country. Efforts will be made to fulfil the goal 'Education for Women's Equality' as laid down in the revised 1992 National Policy on Education (NPE).

Decision-Making: Major Groups Involvement

Considering the strong impact of environmental factors on the sustenance and livelihood of women, full participation of women will be ensured in the conservation and protection of the environment. Further, women will be involved and their perspectives reflected in the policies and programmes of eco-system and natural resources management.

Programmes and Projects

With a view to making women economically independent and self-reliant, a number of interventions have been launched. The Support to Training and Employment Programme (STEP) seeks to train women for employment in the traditional sectors of agriculture, animal husbandry, dairy, handlooms, handicrafts, etc. Launched in 1987, STEP has benefitted more than 250,000 women. A budget of Rs. 160 million has been made available for STEP during the year 1996-97.

The Mahila Samridhi Yojana (MSY), a central sector scheme, was launched on October 2, 1993. This scheme not only inculcates the habit of thrift amongst rural women but also gives them possession and control over their household resources. The scheme has received an overwhelming response from all over the country. Up to September 1996, a total of about 20 million MSY Accounts have been opened with total deposits amounting to over Rs. 2 billion

Economic empowerment of women is mainly based on their participation in decision making processes with regard to raising and distributing resources; that is incomes, investments, and expenditure at all levels. The entire effort of empowering women is to help them exercise their rights in decision making at all levels and in every sphere, within and outside the household, as equal partners in society. Efforts will be made to enhance women's capacity to earn, have access to, and control/ownership of all family/ community assets. In support of women in the informal sector, Rashtriya Mahila Kosh will be further strengthened/ expanded to extend both 'forward' and 'backward' linkages of credit and marketing facilities.

Research and Technologies

Application of science and technology is vital for the advancement of women. Technology will reduce household drudgery and provide better working conditions for women, particularly in rural areas. Emphasis will be given to the improvement of the environment and quality of life of women at an affordable cost.

Cooperation

The Convention on the Elimination of All Forms of Discrimination Against Women was signed by India in 1993.

India endorsed the three priority themes of equality, development, and peace from the Fourth World Conference on Women held in Beijing in 1995, and believes that economic independence and equality in tandem, would create the necessary environment for the realization of the full potential of women. India also endorsed the Commission on the Status of Women as the most appropriate mechanism to fulfil this task. A National mechanism to monitor the implementation of the Platform of Action of the Beijing Conference has been instituted. India believes that National commitments must be complemented by commitments at the international level.

CHILDREN AND YOUTH

Decision-Making: Legislation and Regulations

The Convention on the Rights of the Child ratified by India in 1992 is the guiding principle for formulating policies and programmes of child development. In addition, the existing National Policy For Children (1974) is being suitably reviewed.

India has always followed a pro-active policy for tackling the problems of child labour. The present regime of child labour laws has a pragmatic foundation consistent with the International Labour Conference Resolution of 1979 which called for a combination of prohibitory and mitigating measures. The policy of the Government is to ban employment of children below the age of 14 years in hazardous employments, and to regulate the working conditions of children in other employments. The Child Labour (Prohibition and Regulation) Act, 1986 seeks to achieve this objective.

Decision-Making: Strategies, Policies and Plans

Investment in child development is viewed not only as a desirable societal investment for the nation's future, but also as fulfillment of the rights of every child to 'survival, protection, and development' so as to achieve their full potential.

The thrust in youth affairs has been to involve youth in the entire range of the developmental process recognizing youth as a major resource in the task of nation building. Youth activities will continue to focus on environmental and health programmes such as greening of wastelands, solid waste management, anti-smoking campaigns, prevention of drug abuse, health education with emphasis on reproductive health and prevention of AIDS, population control, and various adventure activities.

The Government announced the National Child Labour Policy (NCLP) in August, 1987. The Action Plan under the Policy comprises: a legislative action plan; focusing of general development programmes to benefit child labour; and a project-based action plan in areas of high concentration of child labour. A major activity undertaken under the NCLP is the establishment of Special Schools to provide basic needs like non-formal education, pre-vocational training, and supplementary nutrition to the children withdrawn from employment.

Programmes and Projects

A significant programme for the development of children during the Eighth Five Year Plan (1992-97) has been the launching of the Integrated Child Development Services (ICDS), one of the world's largest and unique programmes, which aims at providing an integrated package of health, nutrition, and educational services to children below six years, pregnant women, and nursing mothers. ICDS now covers 18.4 million children and 3.8 million mothers who are reached through 3,946 ICDS projects/342,000 Anganwadis. Out of this total, 755 projects are being implemented with the World Bank assistance in the predominantly tribal and backward areas of Andhra Pradesh, Orissa, Bihar, and Madhya Pradesh. In 507 ICDS blocks, services such as health, nutrition, etc. have been extended to nearly 350,000 adolescent girls in the 11-18 year age group, particularly school drop-outs.

Some of the other activities undertaken include: a) adopting the National Plan of Action for children and the girl child; b) adopting the National Nutrition Policy (1993); c) establishing a National Creche Fund for child care services; d) providing creches for children of working/ailing mothers; e) the Balwadi Nutrition Programme; f) early childhood education through assistance to voluntary organizations; g) the Balsevika Training Programme; and h) assisting voluntary organizations in the field of women's welfare and child development. The scheme of day care centres is being implemented through the Central Social Welfare Board in order to provide day care services to children below 5 years and belonging to the weaker sections of society.

During the Ninth Five Year Plan (1997-2002), following the universal application of ICDS and the availability of basic minimum services for the overall development of the child, emphasis will be on consolidation and content enrichment of ICDS through adequate nutrition, supplemented with necessary health check-ups, immunization, and referral services. In this respect, priority will be accorded to the child below 2 years. To achieve this, ICDS will continue to be the mainstay of the Ninth Plan to promote all round development of the young child.

Capacity-building, Education, Training and Awareness-raising

Major schemes undertaken during the Eighth Plan include the National Service Scheme (NSS) and Nehru Yuva Kendras (NYK). The NSS of Youth Affairs provides for the development of their personalities through community services. The programme has successfully incorporated activities which have a social orientation like literacy, environment enrichment, National integration, significance of community management of resources, etc. The NYK scheme aims at providing the rural and non-student youth with opportunities to take part in the process of National development in order to develop their own personality and skills. During 1992-95, the activities conducted include camps, vocational training, rural sports and games, rural cultural activities, youth club development programmes, functional literacy implementation, and a campaign on human survival value.

During the Ninth Plan, greater access will be given to rural and marginalized youth in the vocational training programmes of NYK by involving NGOs, self-help groups, and community polytechnics in the task. The thrust in sports will be on providing greater access to sport facilities through substantial investments in physical education, infrastructure development (including centres of sports physiology and sports medicine), and in creating widespread awareness for physical fitness through nutrition, health education, and yoga with special focus on school children. Area specific sports programmes, recognizing the traditional sports skills of the inhabitants especially tribal populations, will receive priority consideration. Rural sports programmes will be revamped in order to tap the vast talents available in the rural areas. Special attention will be accorded to the promotion of sports and games among the disabled. The need for a holistic approach that will integrate youth programmes within the context of education is well recognized and will guide all actions.

Financing

Voluntary agencies are being financially assisted up to 75% for taking on welfare projects for working children under a Grants-in-aid Scheme.

INDIGENOUS PEOPLE

Decision-Making: Legislation and Regulations

The Constitution of India provides special privileges to Scheduled Tribe communities. The Constitution provides a reservation of 7.5% of vacancies for the Scheduled Tribes in the matters of employment and promotion. Tribal development in the Programme for Special Central Assistance (SCA) for Scheduled Tribes is an additive to State Plans for implementation various socioeconomic programmes for the welfare of Scheduled Tribes.

Decision-Making: Strategies, Policies and Plans

An Action Plan incorporating total food and nutrition security, health coverage, education, and financial assistance in keeping with their socio-cultural conditions is being prepared by the Government. The Action Plan will have flexibility to cater to the specific needs of each Tribe and its environment.

During the Ninth Five Year Plan (1997-2002), a high priority will be accorded to empowering Scheduled Tribes, both economically and socially, to enable them to join the mainstream of National development. To this effect, efforts will be made to create an environment conducive to their being able to lead a life of freedom and dignity, exercising their rights and privileges like any other citizen in the country. The necessary legislative support for this purpose will be provided. The development of the Scheduled Tribes will be consistent with the concept of economic growth with social justice.

The Government will make efforts to minimize the gap that exists between these target groups and the rest of society by all round development, in both qualitative and quantitative terms, and by taking advantage of inputs from both governmental and non-governmental agencies. The Government will also make efforts to ensure that the tribal economy is protected and supported against threats from external markets. The ownership/patent rights of tribal people in respect to minor forest produce vis-a-vis the use of medicinal plants will be protected.

Programmes and Projects

The Centrally sponsored scheme of post-matric scholarship for Scheduled Tribe students has been modified from 1995, thereby revising the maintenance allowance rates, income ceiling for eligibility, and study charges. The restriction of providing benefits of the scheme to two children per family has been relaxed in the case of girl students pursuing correspondence courses and they are now eligible to get a book allowance in addition to earlier reimbursement of non-refundable fees.

Status

The total population of the country is indigenous to India. However, about 7% of the people belong to tribal communities. The Tribal groups are backward both economically and socially. Existing development programmes have not been able to fully alleviate their condition.

The 1991 census shows that most of the Scheduled Tribe population living in rural areas have low levels of literacy and are employed mostly in the primary sector. A Special Component Plan (SCP) and Tribal Sub-Plan (TSP) have been designed to channel the flow of funds from various sectors of development under State/Central Plans to benefit the Scheduled Tribes and their socioeconomic development. These special plans are being monitored in the sectoral plans for education, health, family welfare, housing and urban development, and women and children to ensure better provision for disadvantaged tribal groups.

Financing

A National Scheduled Tribes Finance and Development Corporation has been set up to accelerate economic growth and development for the members of the Scheduled Tribes. The Corporation provides funds at concessional rates for starting projects in, for example, agriculture and allied activities, horticulture, animal husbandry and dairy development, minor irrigation, small industries, trades and services, and transport.

NON-GOVERNMENTAL ORGANIZATIONS

Decision-Making: Strategies, Policies and Plans

In India, the Eighth Five Year Plan has placed emphasis on people's participation and voluntary action in rural development. The role of voluntary agencies has been defined as providing a basis for innovation with new approaches and integrated development, ensuring feedback regarding impact of various programmes, and securing the involvement of local communities, particularly those below the poverty line.

All the programmes and activities in the social sector cannot be implemented by Government alone. The participation of the community and the efforts of voluntary organizations have always had a significant role in this sphere. Efforts have been made to strengthen the involvement of non-governmental and voluntary organizations which reach vast sections of the population, and promote community awareness and participation in various programmes. In many schemes, the various Ministries or Departments have been providing funds to non-governmental organizations for undertaking development activities related to drinking water, health, sanitation, education, and the environment, etc.

The Council for Advancement of People's Action and Rural Technology (CAPART) is the agency providing and assisting voluntary action in rural development. Its funds consist of grants from the Government of India. Programmes of the Ministry of Rural Areas and Employment, like the Employment Assurance Scheme (EAS), Integrated Rural Development Programme (IRDP), and Jawahar Rozgar Yojana (JRY), Development of Women and Children in Rural Areas (DWCRA), Training of Rural Youth for Self Employment (TRYSEM), Accelerated Rural Water Supply, Central Rural Sanitation Programme, are implemented by voluntary agencies through the assistance of CAPART. In addition, CAPART has taken initiatives in promoting a variety of activities for the transfer of technology, people's participation, development of markets for products of rural enterprises, and promotion of other developmental activities and delivery systems in the non-government sector.

To bring CAPART nearer to the people and to ensure closer interaction between it and voluntary organizations at the grassroots level, the functioning of CAPART has been decentralized to six regional centres. It is expected that this decentralization will not only result in improved efficiency and efficacy, but will also be successful in promoting, spreading, and strengthening the role of volunteer organizations in rural development.

LOCAL AUTHORITIES

Decision-Making: Legislation and Regulations

A decentralized approach to planning has been introduced in India through a system of Panchayati Raj and Nagar Palika (local self-governments of urban cities/ towns) institutions. With the enactment of the Constitution

Amendment Act (1992), Panchayati Raj Institutions (PRIs) have been revitalized and a process of democratic decentralization has commenced.

Subsequent to the 73rd Constitutional Amendment Act, State Governments have enacted enabling legislation to provide for elected bodies at the village, intermediate, and district levels, with adequate representation from the weaker sections and women. Almost all the States have constituted Panchayati Raj bodies. The State Governments are required to endow the Panchayats with the power and authority necessary to enable them to function as institutions of self-government, including the responsibility of preparing and implementing plans for economic development and social justice. In the Ninth Five Year Plan, it is expected that the 29 subjects identified in the Eleventh Schedule of the Constitution will be transferred to Panchayati Raj Institutions. Correspondingly, this will involve a transfer of resources. In addition, they will require personnel and administrative support. It is expected that staff engaged in particular works or departments will be transferred along with the work to the Panchayati Raj Institutions.

Decision-Making: Strategies, Policies and Plans

Under the provisions of the 74th Constitution Amendment Act, the Urban Local Bodies/ Municipalities prepare plans for the development of urban areas. The municipalities are the focal institutions for the provision of urban infrastructure and delivery of services and the States should endow them with commensurate functional responsibilities and financial powers.

Under Article 243 (G) of the 73rd Constitutional Amendment Act, the Panchayati Raj Institutions will prepare plans for economic development and social justice. Thus, the core function of the Panchayati Raj Institutions would be planning at the local level through the institution of the District Planning Committees. These Committees will provide the umbrella for the preparation of integrated district development plans. However, certain broad principles will have to be established for assigning a role to each of the three-tiers of government. The actual devolution could be based on the rule that what can be done at a lower level should be done at that level, and not a higher level. The Gramsabha would list out priorities and assist in the selection of beneficiaries for various programmes and schemes. In this way, the aspirations of the people would be articulated. Thereafter, the planning process would follow a bottom up approach with the preparation of village plans which would be incorporated into intermediate level plans, and finally merged into a district plan.

The recommendation of the National Development Council (NDC) that 41% of planning resources be set apart for decentralized planning is expected to be the objective during the Ninth Plan. This could include a proportion as untied funds as 'incentive grants' to match the contribution raised by the Panchayati Raj Institutions. Thereafter, sectoral allocations at the State level should be on the basis of demands made from below by the districts and in keeping with National priorities. In this way, it would be possible to bring about both a vertical and a horizontal integration of resources and services. The Panchayati Raj Institutions would provide an umbrella for the convergence of various sectoral, poverty eradication, and area development programmes at each tier. The vertical integration would be facilitated by an integration of area plans from the village to the State level. This would ensure a synergy between macro-level and micro-level objectives.

A comprehensive and time-bound training policy would be formulated in order to ensure that the Panchayati Raj functionaries are equipped with information regarding various programmes and schemes of the governments, available technologies, and other relevant information which has to be disseminated among the local people. While the Central Government could provide for the training of trainers, the State Governments would have to be responsible for training at the more decentralized levels in keeping with local training requirements.

Capacity-building, Education, Training and Awareness-raising

Awareness building among the people will be given top priority. The government machinery, voluntary organizations, and self-help groups will be involved in the process of advocacy and in organizing the people, especially the poor. Participation of people can be encouraged through beneficiary/functional committees which should be given the responsibility of overseeing the implementation of various programmes. Social audit and transparency in the functioning of the Panchayati Raj Institutions is crucial for the growth and development of these institutions. These will be the important goals of the decentralization strategy during the Ninth Plan.

Financing

While the urban local bodies have a share in the revenue of the States, they will have to be permitted to levy their own taxes at the local level. These could include professional tax, property tax, entertainment tax, and motor vehicle taxes, etc. In addition, they could levy user charges and licence fees wherever feasible. Some of the municipalities in cities could also raise resources from the market by bond issue.

WORKERS AND TRADE UNIONS

Decision-Making: Legislation and Regulations

Labour policy in India derives its philosophy and content from the Directive Principles of State Policy enshrined in the Constitution of India. This has been evolving in response to specific needs of the situation to suit the requirements of planned development and social justice. It has been envisaged that economic growth should not only increase production but also absorb the backlog of unemployed and add a substantial proportion of additional work force.

The Government has implemented many welfare measures for the benefit of workers. In 1995, the Government of India implemented a new pension scheme for workers to replace the family pension scheme which was launched in 1971. The Payment of Gratuity Act was amended in May, 1994, scrapping the eligibility ceiling for its application and enhancing the ceiling on gratuity payment.

Decision-Making: Strategies, Policies and Plans

The labour policy during the Ninth Five Year Plan (1997-2002) will rationalize, simplify, and integrate Labour Laws to bring them in tune with the changing socioeconomic scene. At the same time, the existing legislative framework will be strengthened to protect the interests of labour in the unorganized sector. The following specific steps will be taken:

- In the context of the newly emerging labour market scenario, the role of Employment Exchanges will be reoriented from being mere registration and placement agencies to centres for compilation and dissemination of comprehensive market information, promotion of self-employment, career counselling, and vocational guidance;
- For the improvement of the economic and working conditions of workers in the unorganized sector, a multi-dimensional approach will be adopted with the involvement of voluntary organizations, including schemes for the welfare of the unorganized sector and for bringing about awareness among them of their legislative entitlements;
- An integrated approach will be adopted to rehabilitate bonded labour by pooling resources from a variety of sources;
- Abolition of child labour will be attempted through a multi-pronged approach involving identification and enumeration of child labour and their eventual emancipation;

- Women will be provided with access to education, training, and skill development to enable them to improve their productivity and access to employment, especially new jobs involving technological change. Necessary legislative protection for home workers to protect and safeguard their interests and promote their well being will also be initiated;
- Social security will be provided to workers both in the organized and unorganized sectors. An Integrated Comprehensive Scheme of Social Security will be created over time by having single legislation for all existing social security schemes;
- Occupational health and safety measures will be provided at the work place to improve the overall productivity of the workers;
- Educational and training systems will be reoriented and made more flexible to improve their capability to supply the requisite skills and become responsive to labour market changes;
- Functional autonomy will be granted to training institutions to make them responsive to the changing skill requirements of industries. Training, curricula and equipment, tools, and other infrastructure will be upgraded;
- Regional Vocational Training Institutes for Women will be expanded and further strengthened. Concerted efforts will be made to encourage eligible women trainees to enroll in the Women Industrial Training Institutes (ITIs) and Women Wings of general ITIs through suitable incentives, so that the reservation available to them may be optimally utilized.

Status

The majority of the workforce in India is unorganized in nature, with 80% living in rural areas and 64% engaged in agriculture. Only 15% of the workforce is on regular salaried employment; the remaining 85% is self-employed or employed on casual wages. Access of women to employment compared to men is lower because of their inferior access to education and skill development.

BUSINESS AND INDUSTRY

Decision-Making: Coordinating Bodies

Indian business and industry associations, such as the Federation of Indian Chamber of Commerce and Industries (FICCI) and the Confederation of Indian Industry (CII), work in partnership with the Government, especially with the Ministry of Industry, Ministry of Environment and Forests, and the Ministry of Energy (Power and Non-Conventional Energy Sources) as well as Pollution Control Boards both at the Central and State levels.

Decision-Making: Legislation and Regulations

In recent years, business and industry have made significant efforts towards reducing the impact of industrial activities on the environment, including the investment of considerable resources in the development of environmental management systems and environmentally sound technologies. The concept of green business is results oriented and will have far reaching effect on the Indian environment. The International Standardization Organization has introduced systems of quality control (ISO 9000) and methods for verifying environmental soundness (ISO 14000) of companies. The business community has recognized that in order to stay in business, it will increasingly have to integrate environmental consideration into business strategy and long term planning.

Although some improvement in environmental performance can be expected due to the adoption of a systematic approach, it should be understood that the environmental management system is a tool which enables the organization to achieve and systematically control the level of environmental performance that it sets for itself. The establishment and operation of an environmental management system will not, in itself, necessarily result in an immediate reduction of an adverse environmental impact.

Decision-Making: Strategies, Policies and Plans

In order to adopt best environmental management practices, Indian industry has initiated the development of their corporate environment policies or safety, health and environment policies. Many companies, especially export oriented units, now have written environment policies. The trend for such statements is increasing. The practices essentially flow out of the key elements of such policies, environmental objectives, and targets.

Decision-Making: Major Groups Involvement

Indian business and industry is even promoting the cause of sustainable development beyond the boundaries of their enterprise by participating in initiatives, such as population management, social development and community affairs, rural community development, literacy programmes, and HIV/AIDS awareness programmes. Industry and Government have both started recognizing that implementation of normative measures through environmental laws and standards alone will not serve the cause of sustainable development. There has to be a prudent mix of fiscal and regulatory approaches if cleaner production technologies in Indian industry are to be induced.

Status

Indian industry is today on a fast track of growth, and with the Government's commitment for industrial liberalization the pace of growth is expected to accelerate. It is a big challenge for industry to respond to Government's aspirations, so that the economy improves and the advantages of industrialization are passed on to the people. Industry is also aware of its responsibilities towards the environment and is committed to sustainable development. Compliance levels have gone up which is clearly indicated by the fact that out of 1,551 units identified as highly polluting industries,

1,259 units have provided the requisite pollution control facility. The reasons for this high level of compliance includes pressure from mounting legislation, growing awareness and commitment of industry towards social responsibilities, increasing realization that pollution prevention means good business, and increasing public awareness. Further progress at this stage is not a function of desire or intent but one of technological and other feasibility barriers.

SCIENTIFIC AND TECHNICAL COMMUNITY

Decision-Making: Coordinating Bodies

As part of the Ministry of Science and Technology, the Department of Science and Technology (DST) was established in 1971 to formulate policy statements and guidelines on science and technology. Consistent with wider goals and objectives, various programmes and activities of the DST are aimed at encouraging the Scientific and Technological Community and promoting new areas of science and technology.

A Cabinet Committee on Science and Technology has been established at the apex level to take an overall view of scientific efforts and policy guidelines for the development of science and technology in the country. An Empowered Committee of Secretaries on Science and Technology has also been constituted to implement the recommendations of the Scientific Advisory Committee to the Cabinet.

Decision-Making: Major Groups Involvement

Science and Technology Advisory Committees have been set up in most of the development departments such as Steel, Coal, Mines, Petroleum, and Transport to formulate, implement, and monitor science and technology programmes relevant to the concerned sector. In order to promote science and technology activities at the grass root level, State Science and Technology Councils and Departments have been strengthened and their interaction with various scientific institutions and development departments assured for effective implementation of location specific projects and programmes.

Programmes and Projects

Recognizing the need for accelerating the people's participation in decision making, the National Council for Science and Technology Communication (NCSTC) provides a forum for science and technology aimed at introducing a value system receptive to science and technology among the people at large. Programmes of the DST are geared towards generating employment and entrepreneurial skills for motivating university science graduates to participate actively in the economic growth of the country. Science and Technology Entrepreneurship Parks provide links amongst universities, research laboratories, and industry.

For the socioeconomic development of the rural and urban poor, women, the disadvantaged, Scheduled Castes and Tribal populations, a number of technologies have been developed on carp breeding, seed raising, rain water harvesting, soak pits, water filters, water testing kits, low cost toilets, etc. Efforts have been made through the Science and Technology Entrepreneurship Development Programme to create a number of job opportunities through training and awareness. Several programmes to popularize science like Bharat Jan Gyan Vigyan Jatha, National Children Science Congress, radio and television serials on science themes have also been initiated.

Capacity-building, Education, Training and Awareness-raising

The Government has been a major player in capacity building through increasing research support and has achieved full cooperation of the science and technology community in the decision making process. The financial support for basic research has been more than doubled in the last five years. Capability enhancement through training programmes, contact programmes, and fellowships has been encouraged. The integration of science and technology with socioeconomic development has been initiated and different ministries have set up Science and Technology Advisory Committees (STACs) to identify, formulate, and support science and technology programmes relevant to the concerned sector with the participation of industry.

Research and Technologies

To accelerate the development and application of indigenous technology in production processes, a new fund for technology development and application has been established. The DST also fosters international cooperation in science and technology leading to exchange visits and establishment of special joint centres or projects. Areas which need strengthening are exchange of knowledge and concerns at all levels, and broadening the range of development and environmentally sustainable activities. This would lead to models of joint implementation through requisite cooperation and support.

FARMERS

Decision-Making: Strategies, Policies and Plans

The major thrust of agricultural development programmes in India is to improve the efficiency of the use of scarce natural resources, namely land, water, and energy. This can be achieved only through improved productivity in a cost-effective manner, which alone would increase the welfare of the farmers and agricultural labour. Balanced and integrated use of fertilizers, agricultural credit, institutional development, accelerated investments in agriculture, enhancing the competitiveness of agro-exports, and creation of additional irrigation facilities have been encouraged through various schemes and activities of the Government.

The Eighth Five Year Plan has a major focus on employment generation with the goal of near full employment by the turn of the century. The main objective of the strategy is to generate sufficient job opportunities to absorb unemployed and under-employed persons in agriculture and also to provide jobs to new entrants of the labour force. The agriculture sector provides one of the best avenues for creating greater employment growth.

Programmes and Projects

A scheme entitled "Farmers-Scientists Interaction on Agro-Climatic Zone Basis" has been formulated on a pilot basis. The scheme provides direct feedback from farmers to scientists on problems and constraints in agriculture, while scientists can communicate relevant technological advances to the farming community. This regular system of interaction provides a forum for on the spot identification of field problems and suggestions for remedial measures. Seventeen States are covered by this programme and a proposal for covering other States and Union Territories is under consideration.

The National Cooperative Development Corporation has established its Cooperative Farmers' Service Centre Scheme to provide financial assistance to farmers' service cooperatives. Under this scheme, all types of societies engaged in retail distribution of fertilizer and other agricultural inputs, and non-credit activities are covered by assistance depending upon the requirement. The main objective of the scheme is the development of cooperative societies as effective Farmers' Service Centres for the supply of a wide range of agricultural inputs and also to meet the non-credit needs of farmers.

Capacity-building, Education, Training and Awareness-raising

The multi-tier infrastructure has been created at National, Regional, State, Divisional and District levels to train farmers, farm youth, and farm women. The National Institute of Management has been established at Hyderabad to cater to the needs of extension management. Four Extension Education Institutes have been established on a regional basis to provide training in communication technology and extension methodology. Krishi Vigyan Kendras and Farmers' Training Centres also provide grassroots level training facilities to farmers and farm women. Under the scheme "exchange of farmers within the country", in operation since 1990, opportunities are provided to farmers from less developed areas to tour agriculturally developed areas in groups and observe the progress in agriculture, horticulture, animal husbandry, and allied subjects, so that they can adopt the technology on their own farms.

In order to promote participation, practicing farmers, village youth, and school dropouts are working as focal points for disseminating low cost technology and producing the plant material for conservation measures. Organizing self-help groups to institutionalize people's participation to improve household production systems (mushroom cultivation, sericulture, beekeeping, etc.) is emphasized.

Training and demonstrations have been undertaken to disseminate integrated pest management (IPM) technology. Bio-pesticides, like neem-based formulations, are being encouraged through Farmers' Field Schools. A Central Sector Scheme for Women in Agriculture has been launched in seven States during the Eighth Five Year Plan to motivate and mobilize farm women into groups so that the agricultural support, such as technology and extension, can be channelled through them.

SCIENCE

Decision-Making: Coordinating Bodies

The promotion of science and technology for development has been one of the guiding principles of planned development in independent India. The Ministry of Science and Technology with its associate Departments of Biotechnology and Industrial Research; and the Departments of Electronics, Ocean Development, Non-Conventional Energy Sources, Space, and Atomic Energy all have different responsibilities in this field.

The Ministry of Environment and Forests is the nodal agency in the Government for environmental protection and for natural resource management in the widest sense. The linkages with scientific research and development are clear. The Ministry of Science and Technology promotes research in emerging areas, contributes to technology development, provides linkages for future commercialization, gives priority to areas for scientific research, and focuses on programmes based on developmental needs. The Ministry of Environment and Forests participates in international research programmes and coordinates National research in earth and atmospheric sciences, medium range weather forecasting, etc.

Decision-Making: Strategies, Policies and Plans

During the Eighth Five Year Plan (1992-97), the major thrust areas have been basic research in front line fields, innovative research to achieve self-reliance, diffusion of appropriate technologies, and integration of science and technology in socioeconomic and rural sectors. The Council for Scientific and Industrial Research (CSIR) has made significant achievements in the areas of drugs, pesticides, chemicals, biotechnology, etc. The mechanism for the export of technologies and the systems of patenting have been strengthened. A future thrust would be the modernization of various CSIR laboratories, upscaling technologies, and extension of societal programmes.

Decision-Making: Major Groups Involvement

It is also important to note the activities of institutions such as the National Institute of Oceanography, the National Geophysical Research Institute, and the large number of laboratories under the Council of Scientific and Industrial Research (CSIR), most notably the National Environmental Engineering Research Institute (NEERI). Several universities have departments of environmental sciences. There are also major non-governmental organizations involved in science, such as the Centre for Science and Environment (CSE), Tata

Energy Research Institute (TERI), New Delhi, which work in the field of science, environment, and development.

Programmes and Projects

In the context of the New Economic Policy, the steps taken to re-orient science and technology activities include: the creation of a Technology Development Fund, closer interactions with user industries for technology transfer and the launching of application- oriented R&D programmes. Some additional steps are needed, such as vigorous market-oriented research, and creation of a corpus fund from 2-3% of the turnover of major industries for the promotion of industrial R&D. Such a fund would reduce dependence on the budget support from the Government. Awareness is also required, particularly on intellectual property rights and patents, among science and technology institutions and universities in preparation for the post-General Agreement on Tariffs and Trade (GATT) scenario.

Major activities in the science and technology sector include:

- support to research and development projects, National facilities, special technology development programmes, launching of technology mission-mode projects on sugar production technologies, advanced composites, and fly ash utilization and disposal, promoting technology information system, home grown technologies through the Technology Information Forecasting and Assessment Council (TIFAC);
- international science and technology cooperation and joint programmes with developed countries;
- development of technologies for the socioeconomic sector largely directed towards the rural and urban poor; and
- augmentation of facilities for meteorological forecasting, seismological observations, etc.

Information

In the area of oceans, emphasis has been placed on stabilizing the Antarctic and polymetallic nodules programmes, and the development of ocean data and information system. Besides the expeditions to Antarctica, several achievements have been made in the polymetallic nodules programmes, the coastal ocean monitoring and prediction system, the marine satellite information system, the preparation and dissemination of potential fishing zones, and the establishment of a new institute, the National Institute of Ocean Technology (NIOT).

Research and Technologies

Under Biotechnology, the thrust has been in R&D product development, technology transfer and demonstration, integrated manpower development, augmentation of infrastructure facilities and their optimal utilization, special programmes for specific groups and weaker sections, etc. Significant achievements have been made, besides launching of Technology Mission-mode projects, on bio-fertilizers, biological pest control, and aquaculture. There is a need to formulate a biotechnology profile for the country as well as to ensure transfer of technology.

A technology development fund has been created to accelerate the commercialization of indigenous technologies. In future, the emphasis will be to strengthen R&D efforts further, transfer of the knowledge to industry, strengthen international science and technology cooperation, implement the National Centre for Medium Range Weather Forecasting, selective modernization of the infrastructure facilities of aided scientific institutions, etc. The need in this area is the application of research results for technology development leading, through the involvement of industries and users, to improvements in the quality of life. There is a need for the science and technology entrepreneurship development programmes to be linked with the employment generation programmes.

The thrust of space science and technology has been the development and implementation of indigenous satellites and launch vehicles. Significant achievements include the launching of multi-purpose communication satellites; the development of capabilities for the Augmented Satellite

Launch Vehicle (ASLV), the Polar Satellite Launch Vehicle (PSLV), and the Geosynchronous Satellite Launch Vehicle (GSLV); remote sensing applications for forest mapping, crop inventory, ground water targeting, and flood mapping; and integrated management for sustainable development through micro-level planning, etc. In the future, second generation multi-purpose communication satellites will be launched. There is a need for action in indigenous technological development of strategic items, capability for launching India's INSAT class satellites, building the necessary inventories by involving industries, and stockpiling the inventories for future INSAT systems.

INFORMATION

Decision-Making: Coordinating Bodies

Various Ministries of Government of India are responsible for decision-making in the subject-areas allocated to them. In so far as environmental matters are concerned, the Ministry of Environment & Forests is the nodal agency for decision-making on environment-related matters at the national level as well as for dissemination of information to the users in its allotted field.

As stated above, Ministries collect the information in their specific area and submit it to the Government, with suitable recommendations for making a decision on behalf of the Government. The various departments publish their annual reports and periodically bring out brochures and host available information on the current policy and legislation on the internet.

Information management is carried out by each local government through its Department of Planning/Statistics. Each State is responsible for collection, collation, retrieval and dissemination of information on the subjects related to the particular State Government. Such information is and collated from the State Government by the respective Central Ministry in the Central Government according to the subject. The Ministry of Environment & Forests, through its agencies, coordinates information on forest cover, control of pollution in the State and environmental issues.

Decision-Making: Legislation and Regulations

No Regulations/laws have been made so far by the Central Government. However, the matter is being considered in the Ministry of Information Technology.

The Ministry of Environment & Forests has set up a National Environmental Information System (ENVIS) as a decentralized network for collecting, collating, storing, retrieving and disseminating information in the field of environment and its associated areas.

Environmental Information systems:

Since environment for sustainable development is a broad-ranging, multi-disciplinary subject, a comprehensive information system on environment has necessitated effective participation of institutions/ organisations in the country that are actively engaged in work relating to different subject areas of environment. Environmental Information System (ENVIS) has developed a Home-page of the Ministry. It can be browsed on the Internet at URL:http://www.nic.in/envfor/envis.

Information requested by the Central Government is provided on environmental aspects and its related fields. Similarly, each Central Ministry is having its own instrumental measure for collecting information and disseminating it as and when needed.

A set of environmental indicators has been designed for collecting information in environment and its associated fields.

Decision-Making: Strategies, Policies and Plans

The Ministry of Environment & Forests, apart from the ENVIS programme, mentioned in above is also implementing a Sustainable Development Networking Programme (SDNP) as an externally aided project, to provide information on various thematic areas ranging from pollution, biodiversity, wildlife conservation to agriculture, biotechnology, poverty and climate-change. The SDNP also disseminates knowledge on sustainable development and acts as a distributing clearing house of information and functions in close association with ENVIS

SDNP has developed a website to provide linkages with various national and international information systems and could be browsed in URL: *http://sdnp.delhi.nic.in.*

Decision-Making: Major Groups Involvement

Various steps have been taken to develop information network capabilities of both the public and private sectors. The information required in the existing network is fed by the scientists, local authorities, NGOs and the Village Panchayats at the grassroot level.

Private sector contributes information both in the national network as well as local network. The ENVIS and SDNP networks located in the Ministry of Environment and Forests have set up nodes in various private sector organisations for collecting information on several thematic areas related to environment.

Several groups are consulted in the development of an information system like before setting up of SDNP. A Task Force was constituted with individual experts and private sector organizations. The observations/suggestions given by such groups have been incorporated in designing the system.

Programmes and Projects

Periodic review of existing programmes is being taken up and necessary improvement will be made.

ENVIS is a decentralized information system network, having 25 nodes, known as ENVIS centres, located throughout India. Each node has been assigned a specific subject-area on environment for developing a database and has been assigned the responsibility for collection of data on a regular basis in its specific subject-area, which is carried out through a survey, collection from secondary sources, through research results, and so on.

The information network in a country like India, exists at all levels, particularly at the local and grass root level. In the rural areas of the country, Panchayats, Village Protection Committees have been set up to collect information on various aspects at the grassroot level for onward transmission to the local and national level. With internet reaching all districts in India the information collection and dissemination process has improved.

The Government has taken the step to strengthen electronic networking capabilities through its communication network. Ministry of communication, through its existing mechanisms, is providing support for developing internet facilities and communication linkages to all information systems in the country. The National Informatics Centre has also set up District Information Centres in various State. The question of bandwidth and regular supply of power are being addressed.

Within the ambit of the Technology Information Forecasting and Assessment Council (TIFAC), an autonomous body set up by the Government, support has been provided to establish databases on energy and environmental technologies at the Tata Energy Research Institute (TERI), New Delhi and the National Chemical Laboratory (NCL), Pune respectively. The National Environmental Engineering Research Institute (NEERI), Nagpur has a project on the preparation of an information package on cleaner technologies of industrial production. The experience gained in the establishment of information networks and databases, including manpower training and skills development to handle these systems, is an important strength.

Some of the studies completed with TIFAC support include areas like human settlements, industrial raw water treatment, industrial waste water treatment, water treatment technologies, technologies for disposal of thermal power station fly ash, energy conservation technologies (cement industry), energy saving technologies, biotechnology for waste water treatment, and technologies for the treatment of molasses from distillery effluents.

With a well developed space programme, India is fully capable of collating, collecting, analyzing, and applying remote sensing data obtained through its own and international satellites. Some of the uses of remote sensing technology are already far advanced in India and include: a) the efforts of the National Bureau of Soil Survey and Land Use Planning to use remote sensing techniques for the development of geographic information systems (GIS) for soils, topography, and underground water resources; b) the programme of the Forest Survey of India for using remote sensing to determine and monitor forest cover and its status; and c) the programme of the National Wastelands Development Board to map actual and potential wastelands in 146 districts which have more than 15% of their areas under wastelands.

Apart from remote sensing State Governments and the Government of India collect other kinds of data. These include: a) ground surveys of land resources through assessment of the physico-chemical properties of soil along with topography which also leads to the production of useful and necessary village level maps; b) a mammoth project on water resources in the country undertaken by the Central Water Commission; c) the Agriculture Census and the Agriculture Input Survey which provide information on the classification of land, land use, and the levels of application of inputs such as fertilizers and organic manure; d) State data compiled by the Ministry of Rural Development on some of the basic rural indicators like the establishment of biogas plants which provide sources of alternative energy; e) traditional collection of data by the Registrar General of India through the decennial census; and f) collection of data on health, mortality and morbidity indices, as well as data on education, food security, employment, and earnings for the Physical Quality of Life Indices (PQLI).

In a large and populous country like India, collection of data based on complete sampling is not fully relevant in all cases. The expenditure of both resources and time in the collection of such data might decrease its utility for effective decision-making. Hence, a National Sample Survey Organization (NSSO) has been established to provide trends and useful indicators for decision-making. Recently, the NSSO has commenced data collection on a gender-desegregated basis to provide information about women's status. This facilitates the development of mechanisms for improving their role as primary decision-makers and implementors.

Some of the data generated is included as part of a GIS developed by the National Informatics Centre under the General Information Service Terminal-National Informatics Centre (GISTNIC) programme. The database is, however, yet to be fully synthesized and is presently available only to Government users. India has a reasonably well developed informatics network with computers in both Government, and the private and public sectors. These networks are connected through High Speed Optical Cables within India and with international networks. Connections through satellites for very high speed data transmission have also become a reality and India is able to effectively use the data that may be available on international networks. India has online access to various international databases and networks through DIOLOG, STN, EASYNET, ESA-IRS, etc. CD-ROM databases are also available from vendors like DIOLOG, UMI, SILVERPLATTER and others.

Status

ENVIS network consists of a Focal Point, located in the Ministry and 25 nodes, known as ENVIS Centres, located throughout the country in the potential organizations/ institutions at national level.

The SDNP network is also a decentralized information network concept and besides the Focal Point in the Ministry, has set up nodes on thematic areas of sustainable development in various parts of the country. Each node is in the process of developing a website on its ear-marked area with an interface to the local language.

Information in the field of environment and sustainable development is accessible through internet.

A set of indicators has been designed for collection of information through ENVIS and SDNP. For e.g., the contents of oxides of sulphur, nitrogen and suspended particulate matters in case of collection of air pollution data, the BOD level, COD level, OD level, coliform level and total suspended solids are indicators in case of water pollution, and so on. These indicators help in the assessment of pollution level and formulation of research programmes.

Challenges

The various thematic areas which require attention are:

- Conservation programmes of mangroves, wetlands, estuaries, etc.;
- Biodiversity conservation;
- Control of pollution in air, water including marine;
- Hazardous substances management;
- Development of eco-friendly products;
- Coastal zone regulations and managements;
- Joint Forest Management;
- Documentation of medicinal plants, both traditional and herbarium based; and
- Development of eco-cities and eco-villages, etc.

The user-groups like decision-makers, scientists, environmentalists are at the first level, researchers at the second level and the general public at the tertiary level.

Major challenges experienced in implementation are the requirement of financial support to the various nodes for capacity building and infrastructure development

Capacity-building, Education, Training and Awareness-raising

Several measures have been taken to publicize the existence of ENVIS and SDNP, for its use through electronic media and print media.

The existing nodes of ENVIS and SDNP have been financed suitably to strengthen their capability in developing the web site as well as in the collection, collation and dissemination of information to the users.

Information

The suitable information in so far as the Ministry of Environment and Forests is concerned, can be tapped in the Ministry's home.

INDIAN REMOTE SENSING SATELLITE SYSTEM

The Indian Remote Sensing (IRS) satellites are the mainstay of National Natural Resources Management System (NNRMS), for which Department of Space (DOS) is the nodal agency, providing operational remote sensing data services The IRS system has been further enhanced by IRS-1C, IRS-P3 and IRS-1D, the last two having been launched by India's own launch vehicle, PSLV, IRS-1C, launched on 28 December 1995 and IRS-1D launched by PSLV on 29 September 1997, have enhanced capabilities in terms of spatial resolution, additional spectral bands, stereoscopic imaging, wide field coverage and a more frequent revisit capability than its predecessors.

Remote sensing applications in the country, under the umbrella of NNRMS, now cover diverse fields such as crop acreage and yield estimation, drought warning and assessment, flood control and damage assessment, land use/ land cover information, agro climatic planning, wasteland management, water resources management, under-ground water exploration, prediction of snow-melt run-off, management of watersheds and command areas, fisheries development, urban development, mineral prospecting, forest resources survey, etc. Active involvement of the user ministries/departments has ensured an effective harnessing of the potential of space-based remote sensing. An important application of IRS data is in the Integrated Mission for Sustainable Development (IMSD) initiated in 1992. IMSD, under which 175 districts have been identified, aims at generating locale-specific action plans for sustainable development.

SURVEY OF INDIA

Survey of India (SOI), a national survey and mapping organization under the Ministry of Science & Technology fulfils the ever-growing demand of vast variety of maps of the country. In addition to topographical mapping, SOI is also charged with the responsibilities of other related activities, such as R&D programmes in the field of geodesy, geophysical studies, seismis city, glaciology, indigenisation of instruments/equipments, etc. SOI has started creation of Digital Cartographic Data Base of topographical maps on 1:25 K, 1:50 K and 1:250 K scales. The digital data is being used by various agencies for planning and GIS applications. It also undertakes large scale surveys for various developmental projects including hydro-electric, irrigation, command area, canal area, cantt. area schemes. Coastal mapping has also been undertaken in a phased manner to study the effect of submergence due to rise in sea-level and other natural phenomenon. Survey of India also provides support to neighboring countries in the field of survey education, transfer of technology and various other surveying technologies under bilateral programmes. Under Indo-Bhutan Survey Collaboration project, experts in the field of geodesy, cartography, digital cartography, computer application, printing, etc., have been deputed to Bhutan for acceptability tests of various machines/equipment including training to the officers of Survey of Bhutan. A Geomatics Centre has been planned at Delhi to meet the requirement of geomatics, viz., regional and urban planning, resource management, infrastructure development, environmental monitoring, agriculture, irrigation, soil conservation, forestry, railways, airways, inland water transport, mineral resources, etc.

In the recent years Survey of India has been involved in various inter-disciplinary scientific projects like Sea-Level, Modeling and Monitoring (SELMAM) project of Department of Ocean Development (DOD), modernization of cadastral surveys, glaciology programme of DST, etc. Survey of India has been participating in Indian Scientific expeditions to Antarctica and has strong international linkages especially with International Cartographic Association (ICA), International Society of Photogrammetry and Remote Sensing (ISPRS), South Asian Association for Regional Co-operation (SAARC), International Association of Geodesy (IAG), etc. A map awareness drive has been launched by Survey of India to bring attractive maps in convenient folded size and reasonably priced on various themes, viz,. Antique Map Services, Discover India Series, State Map Series, District Planning Map Series, Tourist Map Series, Trekking Map Series, etc. A Survey Training Institute established under UNDP assistance is a premier institution for training in various disciplines of surveying to the trainees sponsored by the department, other State/Central government organisations and neighboring countries.

NATIONAL ATLAS AND THEMATIC MAPPING ORGANISATION

While Survey of India meets the national needs in cartography, some specialized thematic maps required to meet the needs of the specific users are taken care of by the National Atlas and Thematic Mapping Organization (NATMO), operating under the Department. It also concentrates its attention in a number of areas to integrate resource maps with other relevant socio-economic data and represent them in spatial forms, useful for developmental planning. NATMO is trying to develop the new technology of reverse printing for NATMO maps on experimental basis.

It is also trying to introduce the technique of using metallic colors in map printing. These facilities are also being modernized. The new technologies like Remote Sensing Application, GIS, Application of Local Area Network and Wide Area Network are being introduced in this regard.

SURVEY OF NATURAL RESOURCES

The Botanical Survey of India (BSI), established in 1980, is responsible for surveying and identifying the plant resources of the country. With its headquarters at Kolkatta and nine circles located in different regions of the country, the BSI undertakes exploration tours of the country regularly and the results of such tours are published in the form of national, state and district flora. Established in 1916, the Zoological Survey of India (ZSI) is responsible for carrying out surveys of the faunal resources of the country. While the headquarters of the Zoological Survey of India is at Calcutta, it has 16 regional stations located in different parts of the country. ZSI also undertakes regular faunistic survey tours of the entire country.

The Forest Survey of India (FSI), established in 1981, is entrusted with the task of surveying the forest resources of the country. Besides the headquarters at Dehra Dun, FSI has four regional offices located at Bangalore, Calcutta, Nagpur and Shimla. It prepares thematic maps on 1:50,000 scale and forest vegetation maps on 1:2,50,000 scale of the country. The thematic maps are prepared for the entire country on a ten-year cycle. The vegetation represents 19.27 per cent of geographic area. Out of this, dense forest (crown density more than 40 per cent) accounts for 11 per cent, open forest (crown density 10-40 per cent) represents eight percent, while mangrove forest occupies 0.15 per cent.

GIS model and other modern techniques are being used for data management and assessment along with ground truth verification.

- Satellite based Communication Network;
- Application of Remote Sensing Technology with high resolution;

- Suitable high-speed server for easy and quick dissemination; and
- Lap-top facilities at remote locations.

The National Information System for Science and Technology (NISSAT) facilitates coordination of information services in the country. It has helped establish National information centres in various sectors like leather technology, food technology, machine tools and production, drugs and pharmaceuticals, textiles, chemical and allied industries, advanced ceramics, etc.

Financing

Various Ministries utilize the amount for information management and dissemination out of the budget allocated to them. In case of Ministry of Environment and Forests, roughly, 4% of its allocated budget is used by ENVIS.

Efforts are being made to enhance the budget in case of ENVIS and to strengthen the system network depending upon the availability. Efforts are also being made to make SDNP network as a society for getting financial support from the private sector as well as from other corporate bodies to maintain its sustainability.

Cooperation

In case of SDNP, the entire requirement comes from the external sources such as the UNDP and IDRC. The ENVIS network is also linked with INFOTERRA, the Global Information Network of UNEP, in gathering and sharing information on sustainable development. The SDNP is also linked with SDNP-New York in this regard.

ENVIS is publishing a quarterly newsletter 'ENVIRO NEWS', on a monthly basis. The ENVIS has been entrusted with the responsibility of implementing the UNDP and IDRC-assisted project, 'Sustainable Development Network Programme (SDNP)'. A website for SDNP has been set up for accessing information by a wide cross section of users. (http://sdnp.delhi.nic.in).

Similarly the Sustainable Development Networking Programme (SDNP) is a UNDP/IDRC initiative launched world-wide in 1990 to make relevant information on sustainable development readily available to decision-makers responsible for planning sustainable development strategies. SDNP-India is being implemented by the ENVIS a GOI programme, over a period of three years.

Guidelines and assistance in terms of technology and methodology is received from time to time from International Information Networks like INFOTERRA, IRPTC and GEMS of UNEP.

National indicators in the field of environment have been developed on issues related to sustainable development in the existing information network. Information is also retrieved from various other international networks through linkages for providing to policy planners and decision makers.

However, cooperation in other countries at the international level is still required in accessing information on various issues related to sustainable development.

The major issues not covered earlier in this field are as follows:

- Human resource development and gaining of knowledge through training and workshops at international level for various information networks.
- Cooperation on exchange of information from developed and developing countries.
- Capacity building through bilateral cooperation for the existing information networks.

INTERNATIONAL LAW

Decision-Making: Legislation and Regulations

Consistent with National goals and objectives and using the development planning process as a framework, activities and programmes have been initiated by the Government in the context of Agenda 21. These include legislation to enforce environmental protection, especially in the areas of environmental impact assessment (EIA), pollution control, hazardous waste management, and biodiversity conservation.

The following representative cases decided by the Indian judiciary (Supreme Court of India and High Courts) in recent years illustrate the importance given to environmental protection by the Indian legal system:

1. In Mathew Lukose vs. Kerala State Pollution Control Board, the right to a healthy environment is referred to as one of the fundamental rights. This case also addresses the issues relating to competing claims, that is the growth of industries and the definition of the outer limit of pollution so that it does not infringe the citizens right to a healthy environment. The court noted that: "When the degree of pollution crosses the tolerance limits, it invades the rights... and it cannot pass the mustering right of the Constitution". While considering these competing claims the Court also considered the feasibility of an environment audit preceding the licensing of an industry. It also mooted the creation of a National Environment Agency with powers in areas of planning, enforcement, and sanctions. It further noted that: "An institutional perspective must prevail in these areas and related questions must be upgraded to concerns of National priority". Thus, as "The world belongs to us, but we owe a duty, to the posterity, to the unborn, to leave this world at least as beautiful as we found it". [1990 (2) Kerala Law Journal, page 717]
2. In M.C. Mehta vs. Union of India, the construction of the common Effluent Treatment Plants (CETPs) in the 28 industrial areas in Delhi was addressed. The Court also considered issues relating to the expenditure involved in constructing these treatment plants. Considering these expenditures, the Court took the view that the industries

for whose benefit the treatment plants were installed are bound to cooperate. [1996 (2) Scale (SP) Page 89]

3. In Subhash Kumar vs. State of Bihar and Others, the Supreme Court laid down the following Constitutional norms for controlling pollution: a) right to life is a fundamental right under Article 21 of the Constitution and it includes the right of enjoyment of pollution-free water and air for full enjoyment of life; b) if anything endangers or impairs that quality of life in derogation of laws, a citizen has the right to have recourse under Article 32 of the Constitution for removing the pollution of water or air; and c) a petition under Article 32 for the prevention of pollution is affirmable at the instance of affected persons or even by a group of social workers or journalists. [(1991) 1 SCC page 598]
4. In M.C. Mehta vs. Union of India, the relocation of industries located near the Taj Mahal, which is known as "Taj Trapezium", to preserve the world famous monument was discussed. The Court requested a report from the National Environmental Engineering Research Institute (NEERI) to help it decide whether it is necessary to relocate the various industries. Considering the report from the NEERI, the Court outlined an elaborate plan to relocate the industries. Subsequent to the implementation of this decision, the Court sought to consider the following issues: a) Whether all the industries operating in the Taj Trapezium are to be relocated irrespective of the nature of the industry that is hazardous/ noxious/polluting/non-polluting; b) If the answer to the above question is negative, then which type of industries are to be relocated; c) Whether the Government of India agrees with the suggestion of the relocation outside the Taj Trapezium; and d) Relocation scheme may be indicated. The industrial estates, within or outside the Taj Trapezium, where the industries can be shifted, may be indicated. [1996 (3) Scale (SP) page 58]

5. In Indian Council for Enviro-Legal Action, etc. vs. Union of India and Others, the Court sought to frame special procedure for setting up chemical industries. The court stated, "The Central Government shall consider whether it would not be appropriate in the light of the experience gained, that chemical industries are treated as a category apart. No distinction should be made in this behalf as between a large scale industry and a small scale industry or medium scale industry. All chemical industries, whether big or small, should be allowed to be established only after taking into considerations all the environmental aspects and their functioning should be monitored closely to ensure that they do not pollute the environment around them. It appears that most of these industries are water-intensive industries. If so, the advisability of allowing the establishment of these industries in arid areas may also require examination." [1996 (2) Scale 44, page 73]

Status

International law derives from a number of sources, principally international conventions or treaties, international customary law, and the general principles of law recognized by States. In recent years, each of these sources has displayed features of interest to international environmental law. The inherent reservation notwithstanding, treaties and conventions have made a major contribution to developing international environmental law over the last few years.

Challenges

In order to achieve sustainable development, it is imperative to address on a priority basis the principal social, economic, and environmental challenges contained in Agenda 21. Most international agreements are sector specific in nature, concluded at different times with uneven international knowledge and concern. Therefore, innovative approaches are required in the field of progressive development of international environmental law.

India has become a party to international conventions which contribute to environmental protection and sustainable development. India has ratified almost all multilateral environmental conventions including the recent Framework Convention on Climate Change (FCCC), the Convention on Biodiversity, the Convention on Straddling Fish and Highly Migratory Fish Stocks, and the Convention to Combat Desertification in Countries Experiencing Drought and/or Desertification Particularly in Africa. Concrete actions have been taken to meet international obligations under these conventions to reaffirm India's commitment to pursue activities leading to sustainable development.

SCIENCE: ENVIRONMENT: GLOBAL CHANGE: ACTIVISTS AND NGOs

- *CANA - Climate Action Network Australia* - An alliance of Australian regional, state and national environmental, health, community development, and research groups focusing on the problem of climate change.
- *Center for the Study of Carbon Dioxide and Global Change* - On effects of increasing atmospheric CO_2 levels on the biosphere, with emphasis on studies that suggest a beneficial impact from rising CO_2. Also educational materials describing how to conduct CO_2 enrichment and depletion experiments on plants, and weekly data from current experiments by S.B. Idso and others.
- ***Climate Change and Wildlife - National Wildlife Federation*** - NWF works to protect wildlife from the impacts of climate change.
- ***Climate Concern UK*** - Provides information and co-ordination to help individuals reduce energy consumption, to form local groups to organise campaigning for mutual encouragement in achieving energy savings, and to keep up-to-date with climate change effects and research.
- ***Climate Policy Center*** - An independent, bipartisan, non-profit organization. Seeks to develop and advocate politically realistic U.S. greenhouse gas emission reduction policies.
- ***Climate Strategies*** - An international network organization, coordinated from the Centre for European Policy Studies in Brussels, supported initially by the

SHELL Foundation. It is intended to support cohesive policy responses to climate change.

- ***David Suzuki Foundation: Climate Change*** - The David Suzuki Foundation is a science-based Canadian environmental organization, specializing in global warming, sustainable forestry, salmon farming and fisheries.
- ***ECF - European Climate Forum*** - A mechanism to bring together representatives of different parties concerned with the climate problem: industry, technology, major energy users, insurance and finance sectors, environmental NGOs, and scientific experts investigating climate change and options for sustainable development.
- ***Environmental Defence Society Climate Change Page*** - Discussions of greenhouse warming, the UN Framework Convention on Climate Change, and the Kyoto Protocol from a New Zealand perspective. (EDS is a New Zealand NGO.)
- ***Friends of the Earth: Climate*** - Information and links, especially for activists.
- ***Global Dimming*** - Global Dimming the environmental phenomenon effecting weather and climate change. Particles in the atmosphere cause less sun light to reach the earths surface, the opposite of global warming.
- ***Global Warming*** - Global warming is going to be a big problem, say scientists. Planet for life examines the science behind the predictions.
- ***Global Warming*** - Both domestically and internationally, Environmental Defense works to stabilize Earth's climate by reducing the emission of greenhouse gases.
- ***Global warming*** - Public information from the Union of Concerned Scientists
- ***Global Warming and temperature records*** - A global warming skeptic reviews the science of temperature records and other greenhouse issues, carbon sinks, icebergs, and urban heat islands.

- ***Global Warming and the Pacific Islands*** - How global warming is impacting on the Pacific Islands. How Australia's stance on the Kyoto Protocol is affecting our relations with the neighbouring island states. Many useful links!
- ***Global Warming and the USA — doesn't the US care?*** - With the rejection of the Kyoto Treaty the government of the USA has once again shown the world that environmental issues are not important to them. I will try to solve the mystery of why they do not care.
- ***Global Warming FAQ*** - Uses reputable online resources to answer many of the commonly asked questions about climate change, including questions on whether the earth is getting warmer and whether the climate is changing.
- ***Global Warming: Focus on the future*** - Web version of the exhibit "Global Warming: Understanding the Forecast", which was developed jointly by the American Museum of Natural History and the Environmental Defense Fund and which toured the US in 1997.
- ***Greenhouse Warming: Fact, Hypothesis, or Myth?*** - One man's analysis of the data, by Douglas Hoyt
- ***The Greening Issue: The Good Side of Carbon Dioxide*** - Argues that increasing carbon dioxide levels will be beneficial by resulting in increased food production. Calls for research on this topic.
- ***Greenpeace International : Climate*** - Greenpeace has identified global climate change as one of the greatest threats to the planet. Read reports on climate change, follow our campaigns against Arctic and frontier oil development
- ***The Heat Is Online*** - The website version of Ross Gelbspan's 1995 book on global warming science, politics and industry disinformation, is now a website that tracks and updates climate-related developments, including "Extreme Weather Profiles" from 1995-1999; campaigns

of deception and disinformation by the fossil fuel lobby; and "solution" strategies.

- *Its All About Time & Place* - A Critical look at human culture through time, in response to climate change, ecological and societal responses. Chronologies feature Warm/Cold Events and their societal and social consequences.
- *Kyoto Protocol* - Full text of the Kyoto Protocol, along with news and statistics related to the global warming problem.
- *The Lavoisier Group* - Argues that climate change proposals are based on inexact science and would be too expensive for Australia's industry.
- *Lick Global Warming* - A guide to global warming and an interactive map showing its effects. An invitation to pledge to reduce CO_2 emissions and to encourage Congress to take action.
- *NRDC: Global Warming* - NRDC advocates dramatic reduction in emissions of greenhouse gases in the U.S. and abroad as well as the adoption and implementation of the Kyoto Protocol. Includes fact sheets, reports, and links.
- *Safe Climate* - Safe Climate is a source of tools, news, and information for individuals and organizations to take action to combat global climate change. A project of the World Resources Institute.
- *Science & Environmental Policy Project* - Site from Dr. Singer expressing skepticism regarding many aspects of global warming.
- *Sierra Club Global Warming Campaign* - Fact sheets and news items on global warming, its consequences, and measures to reduce its impacts.
- *Still Waiting For Greenhouse* - A lukewarm view of global warming from Tasmania. By John L. Daly.
- *UNDO IT* - A project of Environmental Defense, a US national campaign to ramp up the fight against global

warming, the most critical environmental issue we face. Get the facts, help undo it, and spread the word.

- *What's Wrong With Still Waiting for Greenhouse* - Run by scientists who are concerned about misinformation that appears on the Internet in relation to Global Warming (also called the Greenhouse Effect). In particular, it seeks to expose errors of fact on one particular site - Still Waiting for Greenhouse.
- *WWF - Climate Change Campaign* - Features information and breaking news on climate change, with a focus on resources and actions.
- *WWF's Climate Change Program* - Project to halt global warming by promoting international ratification of the Kyoto climate treaty.

ENVIRONMENT: GLOBAL CHANGE: CARBON CYCLE *(9)*

- **Business: Mining and Drilling: Mineral Exploration and Extraction: Industrial Minerals: Coal** *(13)*
- **Science: Chemistry: Elements: Carbon** *(13)*
- **Science: Technology: Energy: Fossil Fuel: Coal** *(14)*
- *The BOREAS Project* - The Boreal Ecosystem-Atmosphere Study (BOREAS) is a large-scale international interdisciplinary experiment in the northern boreal forests of Canada. Its goal was to improve our understanding of the boreal forests - how they interact with the atmosphere, how much CO2 they can store, and how climate change will affect them.
- *Carbon Dioxide Information Analysis Center (CDIAC)* - Global-change data and information analysis for the US Department of Energy (DOE). Atmospheric concentrations of carbon dioxide and other radioactively active gases; role of the terrestrial biosphere and oceans in biogeochemical cycles of greenhouse gases; emissions of carbon dioxide; long-term climate trends; effects of elevated carbon dioxide on vegetation; vulnerability of coastal areas to rising sea level.

- ***Climate and Carbon Cycle Modeling Group (CCCM)*** - The CCCM is a part of Atmospheric Science Division at Lawrence Livermore National Laboratories. Site provides access to some of their publications, as well as information on the centre, their models and their projects.
- ***Fluxnet Canada*** - A Canadian research network, based on eddy covariance flux measurements, that studies the effects of climate and disturbance on carbon cycling processes in forest and peatland ecosystems.
- ***Global Carbon Project*** - Scientific partnership that aims to develop a complete picture of the global carbon cycle, including both its biophysical and human dimensions together with the interactions and feedbacks between them. A project of the Earth System Science Partnership of the International Geosphere Biosphere Programme (IGBP), World Climate Research Programme (WCRP), and the International Human Dimensions Programme (IHDP)
- ***Mystery of the Missing Carbon*** - Discusses the BOREAS research program, which investigated the "missing carbon" problem in global carbon cycle studies.
- ***NOFRETETE: Nitrogen oxides emissions from European forest ecosystems*** - A Project funded by the European Commission DG Research - Vth Framework programme Programme in the thematic programme "Energy, Environment and Sustainable Development", Key Action 2 "Global change, climate and biodiversity".
- ***United States Trace Gas Network*** - The TRAGNET network is meant to accomplish the following two goals: 1. Document contemporary fluxes of CO_2, CH_4 and N_2O between regionally important ecosystems and the atmosphere and 2. Determine the factors controlling these fluxes and improve our ability to predict future fluxes in response to ecosystem and climate change.
- ***US Carbon Cycle Science Program*** - An interagency partnership to provide critical unbiased scientific information on the fate of carbon dioxide in the

environment to contribute to the ongoing public dialogue.

SCIENCE: ENVIRONMENT: GLOBAL CHANGE: CARBON MANAGEMENT (43)

- **Emissions Trading@** *(40)*
- **Livestock Methane Emissions@** *(3)*
- *Products and Services (8)*
- **Business: Energy and Environment: Consulting** *(542)*
- **Science: Biology: Ecology** *(1,011)*
- **Science: Biology: Microbiology** *(403)*
- **Science: Earth Sciences: Geochemistry** *(67)*
- **Science: Earth Sciences: Oceanography** *(980)*
- *Big Sky Carbon Sequestration Partnership* - The Big Sky Carbon Sequestration Partnership addresses CO2 emissions and seeks the most suitable technologies, regulations, and infrastructure needs for carbon capture, storage and sequestration, both terrestrial and geologic.
- *CarboEurope* - A EU-funded research project designed to better understand, quantify and predict under current and future scenarios the carbon balance of Europe.
- *Carbon Footprint* - Information for householders to identify and calculate their carbon footprint and tips to reduce carbon emissions.
- *The Carbon Fund* - The Carbon Fund is financing and promoting the protection, restoration and enhancement of forests all over the world through carbon sequestration, an important tool in preventing global climate change.
- *Carbon Mitigation Initiative* - Website providing information about a joint research project of Princeton University, BP and the Ford Motor Company into impacts of climate change and measures to reduce greenhouse gas emissions. The site includes research papers that can be downloaded.

- ***Carbon Sequestration*** - Information on US DOE's Office of Science research and development activities in sequestering carbon underground, in terrestrial ecosystems, in the ocean, and through manipulation of microbes.
- ***Carbon Sequestration in Terrestrial Ecosystems (CSiTE)*** - U.S. DOE research aimed at enhancing carbon capture and long-term sequestration in terrestrial ecosystems.
- ***Carbon Sequestration Leadership Forum*** - Information about the CSLF and its activities, including endorsed projects, papers and presentations.
- ***The Carbon Trust*** - The Carbon Trust is an independent not for profit company set up by the UK Government with support from business to take the lead on low carbon technology and innovation in the UK.
- ***Carbon-Dioxide Sequestration in Geologic Media*** - Texas Bureau of Economic Geology describes its research initiatives on the potential to sequester carbon in saline geologic formations and abandoned oil/gas reservoirs.
- ***CELB-Climate Change*** - Carbon offset projects in tropical forests with combined carbon, biodiversity and social benefits.
- ***CIRENE (Center of Initiative and Research on Energy and the Environment)*** - Information about syntheses and studies to solve problems of industrial partners regarding greenhouse gases (including taxation, permits for carbon trading, technical-economic studies, feasibility studies for capture and storage of CO_2, forecasting studies).
- ***Climate Care*** - A scheme that lets individuals and companies pay to offset their CO_2 emissions by investing in renewable energy, energy efficiency and other projects that reduce emissions. Also provides background information and practical advice.
- ***CO2 Capture Project*** - Describes efforts by eight of the world's leading energy companies to reduce the cost of CO_2 capture from combustion sources and to develop

methods for safely storing carbon dioxide underground.

- ***DOE: Carbon Sequestration R&D*** - U.S. Department of Energy program of research and other efforts to develop mid- to long-term solutions for reducing carbon dioxide concentrations in the atmosphere, including sequestration in oceans, geologic formations, and terrestrial ecosystems.
- ***DOE Center for Research on Ocean Carbon Sequestration*** - United States Department of Energy (DOE) view of techniques for storing carbon in the ocean.
- ***Environmental Synergy, Inc.*** - Environmental Synergy, Inc. (ESI) is a service organization providing reforestation and carbon quantification services to corporate clients as a means to offset carbon dioxide (CO_2) emissions and promote sustainable forestry.
- ***ETV Greenhouse Gas Technologies*** - U.S. EPA and Southern Research Institute (SRI partnership provides independent verification of greenhouse gas (GHG) mitigation and monitoring technologies.
- ***Fact Sheet: Soil Carbon Sequestration*** - Frequently asked questions presented by the United States Department of Agriculture (USDA).
- ***Future Forests for a Carbon Neutral World*** - A UK company that will plant trees (for a fee) on behalf individuals and companies in order to offset their carbon emissions.
- ***The GEO-SEQ Project*** - Describes U.S. public-private research and development on methods for capture and geologic sequestration of CO_2.
- ***GHC Protocol Initiative*** - Standards and guidance regarding corporate greenhouse gas accounting and reporting.
- ***Greenhouse Gas Technology Center*** - Investigating the performance of technologies to mitigate and monitor greenhouse gases in the atmosphere.

- ***Hydrogen Production With Carbon Sequestration*** - Research overview of the production of carbon sequestering fertilizer made during renewable hydrogen production.
- ***Land-Use Management for Carbon Sequestration*** - Study of the potential for sequestering carbon from alternative land-use management options in Sub-Saharan Africa developed by Oak Ridge National Laboratory (ORNL).
- ***Midwest Geological Sequestration Consortium*** - The MGSC will assess the potential for geologic sequestration of carbon dioxide in the deeply buried unminable coal seams, depleting oil or gas reservoirs, and brine-filled rock formations found in the Illinois Basin.
- ***MIT Carbon Capture and Sequestration Technologies*** - Background information and details of research projects.
- ***Ocean Chemistry of Greenhouse Gases*** - Information about research by the Monterey Bay Aquarium Research Institute (MBARI) into ocean sequestration of carbon dioxide.
- ***Safe Climate for Business*** - A website dedicated to helping business of all sizes to understand and take action on climate change.
- ***Siberia II - Project Website*** - Multinational study to demonstrate the viability of full carbon accounting on a regional basis using the environmental tools and systems available to us today and in the near future. Site provides news, data and other information.
- ***Sleipner CO_2 project*** - Animation showing the separation process of CO_2 and natural gas as used by Statoil prior to underground CO_2 storage.
- ***US EPA Methane Voluntary Programs*** - EPA methane and utility outreach programs are intended to reduce emissions of methane, a potent greenhouse gas.
- ***U.S. Agriculture and Forestry Greenhouse Gas Inventory: 1990-2001*** - A comprehensive assessment of greenhouse gas emissions and sinks in U.S. agriculture and forests.

Provides extensive, in-depth emissions and sinks estimates for livestock, cropland, and forests, as well as energy consumption in livestock and cropland agriculture (March 1, 2004)

- ***Trends of Measured Climate Forcing Agents*** - Paper by Hansen and Sato (PNAS 2001;98:14778) makes the case for an increased focus on methane and aerosols in the short-term in order to mitigate climate change. (December 18, 2001)
- ***The Royal Society - The role of land carbon sinks in mitigating global climate change*** - A report that highlights the considerable uncertainty in the scientific understanding of the causes, magnitude and permanence of the land carbon sink. The full report (27 pages) and a summary are available. (July, 2001)

Science: ***Environment***: ***Global Change***: Climate Prediction (22)

- ***Science: Earth Sciences: Meteorology: Climate Prediction*** (23)
- ***Hadley Centre for Climate Prediction and Research*** - Provides, for the UK Government, an assessment of both natural and man-made climate change. Offers accessible information on climate modeling and climate change in general, with particular emphasis on data from the Hadley Centre models.
- ***Abrupt Climate Change Research at LDEO of Columbia University*** - An overview of paleoclimatological, observational and modeling evidence to identify causes and impacts of past and future climate changes. Includes a Frequently Asked Questions (FAQ) page.
- ***Australian Bureau of Meteorology Research Centre*** - The research division of the National Meteorological Service in Australia.
- ***Canadian Centre for Climate Modelling and Analysis*** - A division of the Climate Research Branch of Environment Canada's Meteorological Service of Canada. Site provides

an overview of research and information on their models and results.

- *Climate Change Prediction Program* - A U.S. Department of Energy program to rapidly advance the science of decade- and longer-scale climate prediction.
- *Climate Modelling and Applications* - CSIRO's Climate Modelling and Applications Team seeks a better understanding of climate and its variations in order to assess the way in which climate is likely to change in future due to the enhanced greenhouse effect and to natural climatic variability.
- *Climate System Technology (TSI)* - Aims to improve understanding of the Indonesian maritime continent climate variability as an aid natural resource development through cooperative work with both national and international scientific communities
- *Climate Variability and Predictability Study (CLIVAR)* - An interdisciplinary research effort within the World Climate Research Programme (WCRP) focussing on the variability and predictability of the slowly varying components of the climate system. It investigates the physical and dynamical processes in the climate system that occur on seasonal, interannual, decadal and centennial time-scales.
- *ClimatePrediction.net* - Aims to harness the power of PCs in homes and businesses to predict the climate of the 21st century. Individuals can register to take part when the experiment starts (likely sometime 2001).
- *Deutsches Klimarechenzentrum* - German Climate Research Centre, Hamburg. Provides information in English on the Institute and their research.
- *Experimental Climate Prediction Center* - Information and long term predictions from the climate division of the Scripps Institution of Oceanography at the University of California, San Diego.
- *Geophysical Fluid Dynamics Laboratory Home Page* - News, articles, data and other information on climate modelling.

- ***Global Energy and Water Cycle Experiment*** - Observes and models the hydrologic cycle and energy fluxes in the atmosphere, at the land surface, and in the upper oceans. GEWEX is an integrated program ultimately leading to the prediction of global and regional climate change.
- ***Japanese Meteorological Research Institute*** - Provides information on their climate simulations and on their current research.
- ***Laboratoire de Meteorologie Dynamique du CNRS*** - Information (in English) on the centre's climate modelling activities is provided in the 'introduction' link.
- ***Max Planck Institute for Meteorology*** - Information on climate models and research activities.
- ***MIT Center for Global Change Science*** - Conducts research to enhance our ability to accurately predict changes in the global environment. Areas of interest include Convection, Atmospheric Water Vapor, and Cloud Formation; Oceans and Ocean-Atmosphere Coupling; Land Surface Hydrology and Hydrology-Vegetation Coupling; Biogeochemistry of Greenhouse Gases and Reflective Aerosols; and Upper Atmospheric Chemistry and Circulation.
- ***NASA Goddard Institute: Global Climate Modeling*** - Information on climate mdoling studies at the Goddard Institute.
- ***Numerical Terradynamic Simulation Group, University ofMontana*** - Research group focusing on remote sensing and computer simulation of terrestrial ecosystem processes, with active projects at local, regional, continental, and global scales.
- ***Project to Intercompare Regional Climate Simulations (PIRCS)*** - PIRCS is an international program to evaluate strengths and weaknesses of regional climate models by comparing how well these models are able to simulate present-day climate.
- ***UK Universities Global Atmospheric Modelling Programme*** - Aims to strengthen the role of the UK

universities in the vital area of numerical modelling of the large scale atmosphere.

- *US NCAR Climate and Global Dynamics Division* - The National Center for Atmospheric Research's division devoted to climate modelling. Provide newsletters and an overview of their research.

Science: *Environment*: *Global Change*: Desertification (20)

See also:

- *Science: Biology: Ecology: Ecosystems: Deserts* (43)
- *Science: Earth Sciences: Meteorology: Weather Phenomena: Drought* (11)
- *Society: Issues: Environment: Conservation and Endangered Species: Ecosystems and Habitats: Deserts* (1)
- *Desertification* - FAO website on the topic, with technical and scientific data and information available at FAO, plus internet links. Site available in Arabic, English, French and Spanish languages.
- *Adaptive Strategies for Sustainable Livelihoods in Arid and Semi-Arid Lands (ASALs) Project* - Description of ASALs Project in Africa, along with outputs, findings, significance and bibliography.
- *Beijing's Desert Storm* - Article and photos describe desertification in China, where the desert is on the move. With dunes already 75 kilometers from Beijing, environmentalists fear the sand could swallow the capital in a few years.
- *Combatting Desertification* - Introduction to desertification and ways to combat desertification, details of activities undertaken and contact information. A site by Canadian International Development Agency.
- *Desertification and Drought* - Internet links to sites related to desertification, maintained by University of Arizona.
- *Desertification: Monitoring & Forecasting* - Text and video reports about remote monitoring in the arid

southwest region of United States. General details about desertification, forcasting, desert plants and a photo gallery is also there.

- ***Drylands: Bright Edges of the World*** - Smithsonian Institution electronic exhibit about deserts and other arid and sub-humid environments, emphasizing desertification and other degradation and its consequences.
- ***Land Degradation and Desertification*** - International Union of Soil Sciences working group site offers conference reports, technical papers and links to other organizations.
- ***MEDACTION: Policies for Land Use to Combat Desertification*** - Describes a decision support system for land management in Italy and Portugal. Features project description, bibliography and contact details.
- ***Medalus - Programme on Mediterranean Desertification and Land Use*** - Information about international project, funded by the European Union, to investigate the effects of desertification on land use in Mediterranean Europe.
- ***Monitoring and Early Warning*** - Maps providing monthly updated, near real time, distributed information on rainfall, actual evapotranspiration, radiation, drought and desertification indices, and crop yield forecasts for Africa and Europe. Includes methodology discussions and links to similar resources.
- ***Past, Recent, and 21st Century Vegetation Change in the Arid Southwest***
- ***The Sahara and the Sahel Observatory (OSS)*** - Details about mission, objectives and structure of OSS, an organization striving to build up an African arena for cooperation and exchange to combat desertification and poverty. Available in English and French.
- ***Selected Internet Resources on Desertification*** - Information about intergovernmental organizations, national/regional activities, non-governmental

organizations in the field of desertification. Also related reports/news/information sources.

- *UN Convention to Combat Desertification Information Network* - Objective is to provide information services on the subject of desertification. Background on the phenomenon, key international documents, and directory of organizations conducting work in this area.
- *UNDP Dryland Web - Office to Combat Desertification and Drought (UNSO)* - Works in the overall framework of UNDP's sustainable human development mandate to support dryland development and the implementation of the CCD.
- *United Nations Convention to Combat Desertification* - Provides access to the official documents maintained or received by the UNCCD Secretariat and is a general source of information on the topic of desertification for the interested public.
- *Web Resources on Desertification* - Links to web sites providing information on desertification worldwide.
- *The Second Meeting of the Conference of the Parties to the U.N. Convention to Combat Desertification* - Page includes summaries of daily discusses, which featured concerned nations from around the world. Past event took place in Dakar, Senegal. (November 30, 1998)
- *First UN Convention to Combat Desertification* - Purpose of meeting was to discuss the problem of land degradation and to offer suggestions to prevent this phenomena. Page includes list of participating countries and their ideas. Took place 29 September - 10 October 1997, Rome Italy. (September 29, 1997)

SCIENCE: ENVIRONMENT: GLOBAL CHANGE: IMPACTS AND INDICATORS **(82)**

- **Animals** *(15)*
- Plants *(5)*
- **Ecosystems** *(7)*
- Regional *(20)*

- **Science: Earth Sciences: Meteorology: Weather Phenomena** *(501)*
- **Science: Environment: Biodiversity** *(983)*
- **Society: Issues: Environment** *(2,567)*
- *Assessment of Knowledge on Impacts of Climate Change* - A detailed report for the German Advisory Council on Global Change (WBGU) into the effects of dangerous anthropogenic interference on ecosystems, food production, water, and socio-economic systems. Studies include: processes causing loss of biodiversity and ecosystem damage; processes driving species endangerment and extinction; projected effects on species and ecosystems. [PDF]
- *Climate Monitor* - Regular updates of climate and meteorological data, plus press and media commentaries. Includes monthly weather summaries and climate time visualisation, with links to the main data pages of the Climate Research Unit and relevant sites.
- *A Consortium for the Application of Climate Impact Assessments (ACACIA)* - Aims to increase knowledge of the science of climate change and variability. Includes data sets, details of current projects, events, and publications, and links to related web sites.
- *Effects of Climate Change on Protected Areas* - WWF paper that examines recent research and provides evidence that the types of environmental changes predicted in climatic models are now taking place. Studies the implications to habitats, ecosystems, species, and local food webs, in areas that are rooted in the concept of permanence. [PDF]
- *Global Change Data and Information System (GCDIS)* - A collection of distributed information systems operated by government agencies involved in global change research. Includes multidisciplinary data from atmospheric science, ecology, oceanography, economics and sociology.
- *Global Change Master Directory* - Database including descriptions of several Earth science fields: the

atmosphere, biosphere, hydrosphere and oceans, snow and ice, geology and geophysics, paleoclimatology, and human dimensions of global change.

- ***Global Climate Change and Biodiversity*** - Summary report from an international conference presenting papers on topics that cover a cross-section of the planet's major biomes: forests, marine, high latitudes and montane, managed landscapes and coasts. [PDF]
- ***Global Warming and Hurricanes*** - Examines the likely changes in worldwide hurricane activity in the 21st century. Includes graphs and diagrams, links to theoretical papers, and a research report.
- ***Global Warming Facts and Our Future*** - An introduction to the subject, designed for teenagers and adults. Examines the greenhouse effect, carbon cycle, causes and effects, predictions, and responses to change. Includes teaching activities, links, and suggestions for further reading. From the Marian Koshland Science Museum.
- ***Global Warming Map: Early Warning Signs*** - Illustrates observed consequences, as indicated by periods of unusually warm weather, coastal flooding, and changes in glaciers and polar regions.
- ***Gulf Stream Shutdown*** - Abrupt climate change, global warming and the environment. If the Gulf Stream slows down or stops, what could be the effects? Websites, articles, books and forums.
- ***Habitat: Climate Change*** - An overview of the growing evidence of the sensitivity and vulnerability of natural and human systems, including brief descriptions of threats encountered by specific species: golden toad, western toad, salmon and trout, polar bears, arctic fox, chinstrap and adelie penguins, gray wolves, painted turtles, and tree swallows.
- ***Human Dimensions of Global Change Specialty Group*** - The HDGC Specialty Group promotes the interests of geographers who are united by exploration of human

processes that affect or are affected by environmental changes.

- ***Intergovernmental Panel on Climate Change (IPCC) - Working Group 2 : Impacts, Adaptation, and Vulnerability*** - Assesses the scientific, technical, environmental, economic and social aspects of climate change.
- ***Intergovernmental Panel on Climate Change (IPPC) - Working Group 1 : The Scientific Basis of Climate Change*** - Analyses a body of observations of all parts of the climate system; catalogues increasing concentrations of greenhouse gases; assesses the processes and feedbacks which govern the climate system.
- ***International Year of the Ocean: Impacts of Global Climate Change*** - A document prepared by various participating US Government agencies as a background discussion paper for this declaration. Considers the influence of the ocean in three parts: seasonal to interannual effects; decadal to centennial effects; and with an emphasis on US coastal areas.
- ***Long Term Sea Level Change*** - Data and figures providing information on global sea level changes provided by tide gauge data and by the TOPEX/Poseidon satellite. From the Center for Space Research, University of Texas.
- ***National Institute of Water and Atmospheric Research (NIWA) Project FOE03103*** - A modeling case study by Friends of the Earth. Examines and assesses the effects of specific emissions from a particular enterprise, ExxonMobil, on atmospheric concentrations, changes in radiative forcing, changes in global mean surface temperature, and changes in sea level. [PDF]
- ***National Snow and Ice Data Center (NSIDC): State of the Cryosphere*** - Information on the status of sea ice, freshwater ice, snow, glaciers, and permafrost. Includes data sets with search facility, details of projects and research, news articles, and links to data providers.

- ***Ocean and Climate Change Institute: Abrupt Climate Change*** - Woods Hole Oceanographic Institution's study of the role of ocean circulation in climate regulation, and research into changing salt levels.
- ***Office of Satellite Data Processing and Distribution (OSDPD)*** - Graphical presentation from the US Government National Environment Satellite, Data, and Information Service. Includes charts for sea surface temperatures, aerosol optical thickness, coral bleaching hotspots, precipitation, volcanic ash, and a snow and ice mapping system.
- ***Science Daily: Earth and Climate News*** - A daily news update of the latest research into climate change. Includes links to archives articles about ecology and environmental science.
- ***Simulation of Recent Southern Hemisphere Climate Change*** - Observed climate trends simulated in an atmospheric model. Evidence that anthropogenic emissions of ozone-depleting gases have had a distinct effect on climate at Earth's surface as well as at stratospheric levels. [PDF]
- ***Temperature Trends: Atmospheric (Satellite MSU - RSS)*** - One of two conflicting analyses of atmospheric temperatures derived from satellite observations. Browse images and download data. From Remote Sensing Systems, in collaboration with the NOAA Air Resources Laboratory.
- ***Temperature Trends: Atmospheric (Satellite MSU - UHA)*** - One of two conflicting analyses of atmospheric temperatures derived from satellite observations. Interactive tool showing satellite observations of regional and global trends in the lower troposphere and stratosphere. From the Global Hydrology and Climate Center in Huntsville, Alabama.
- ***Temperature Trends: Surface (CRU)*** - Land and sea surface temperature anomalies, analysed by the Climate Research Unit, Norwich, UK. Includes scientific papers,

dataset terminology, file formats, data for downloading, answers to frequently asked questions, and links to related web sites, grids, and datasets.

- *Temperature Trends: Surface (GISS)* - Global Historical Climate Network data, analysed by NASA's Goddard Institute for Space Studies. Includes background information, table data, individual station data, graphs, maps, animations, and references.
- *The Threat of Global Warming* - Provides an broad overview of the subject, including notes on diverse environmental topics and links to brief essays about threatened species and habitats. From the environmental organization EcoBridge.
- *Tropical Cyclones and Global Climate Change* - A Bulletin of the American Meteorological Society. Examines the natural variability, possible trends, frequency, and intensity of severe weather activity in tropical coastal regions, and explores the potential for change in greenhouse conditions.
- *Understanding Climate Change Feedbacks* - An e-book version of the publication from the Board of Atmospheric Sciences and Climate (BASC), produced by the National Academies Press. Examines processes in the climate system that can either amplify or damp the response to changes forced by external factors.
- *World View of Global Warming* - Photographic documentation of change in the Arctic, Antarctica, glaciers, temperate climate zones, and the ocean. Focuses on effects such as shrinking glaciers, coral bleaching, insect and animal range changes, phenology, and rising sea level.
- *Climate Change Threatens a Million Species with Extinction* - Detailed press release reporting a study of biodiversity-rich regions. Computer models project the future movements and distributions of many species of plants, birds, animals, and insects, in response to

changing conditions. From the School of Biology, University of Leeds. (January 7, 2004)

- *NAS: Abrupt Climate Change - Inevitable Surprises* - Detailed report on the mechanisms and possibility of rapid climate change, including the potential link with greenhouse warming. (December 11, 2001)
- *Climate Change Linked to Civilization Collapse* - National Geographic News story about the hypothesis that past climate variations led to the collapse of civilizations, with speculation on the implications for our society's future. (February 27, 2001)
- *Global Sea Level Change: Determination and Interpretation* - Full text of a detailed report published by the American Geophysical Union. (January 1, 1995)

SCIENCE: ENVIRONMENT: GLOBAL CHANGE: *POLICY (35)*

Top: Science: Environment: Global Change: *Policy (35)*

- **Science: Social Sciences: Economics: Environmental Economics** *(75)*
- **Society: Issues: Environment: Climate Change** *(115)*
- *Austrian Council on Climate Change* - Interdisciplinary working group which aims to identify and evaluate measures relating to the prevention of global climate change. English version accessed from left hand menu.
- *BC Climate Exchange* - Provides clearinghouse of resources to make climate change relevant, by showing link to sustainability solutions — like green buildings, community planning and energy efficiency — that also have other social and economic benefits. Coordinated by the Fraser Basin Council.
- *Center for Climate Change and Environmental Forecasting* - U.S. Department of Transportation initiative. Fosters awareness of the potential links between transportation and global climate change, and formulates policies to deal with the challenges of these links.

- ***Center for International Climate and Environmental Research (CICERO)*** - Norwegian Government organisation mandated to conduct research and disseminate information on international climate issues. Site provides news, articles, and in-depth publications.
- ***Choose Climate*** - Interactive graphics link climate science and policy. Adjust the parameters and uncertainties just by dragging a control with the mouse to see effects of various per-capita emissions.
- ***Cities for Climate Protection*** - An innovative program that helps local government and their communities to reduce greenhouse gas emissions and their impact on the environment.
- ***The Clean Development Mechanism: A Primer*** - Informative article about CDM from the Resources for the Future (RFF).
- ***Climadapt Canadian climate change adaptation network*** - A private sector driven environmental network providing innovative climate change adaptation expertise in Canada and internationally
- ***Climate Change - Office of International Information Programs, U.S. Department of State*** - Information on climate change, global warming, U.S. policy, conventions and forums.
- ***Climate change - UNFCCC*** - Information on the UN Framework Convention on Climate Change provided by the International Institute for Sustainable Development.
- ***Climate Change News and Actions by SafeClimate.net*** - Tools, news, and information for individuals and organizations to take action to combat global climate change.
- ***Climate Change Solutions*** - Success stories, education materials, interactive tools and resources on actions to reduce greenhouse gas emissions for individual, municipalities, and industry, particularly in Canada.

- ***Climate Institute*** - An NGO that aims to serve as a bridge between policymakers and scientists around the world and is dedicated to being the world's foremost authority on climate change information, science and responses. Site provides basic scientific information and links to more detailed information.
- ***Climate Solutions - Practical Solutions to Global Warming*** - Mission is to stop global warming at the earliest point possible by helping the Northwest become a world leader in practical and profitable solutions.
- ***Cool Kids For A Cool Climate*** - Explanation of global warming, causes and effects, kids' stories, ways to take action, links, and news.
- ***CorpWatch - Climate Justice Initiative*** - Articles and background information from an NGO that campaigns against multinational corporations.
- ***Forum for the Study of Crisis in the 21st Century*** - To analyse the nature of humankind's current crisis. Bringing together diverse academic and independent researchers in an independent research-based centre. Current programmes; climate change and violence, climate change and public opinion.
- ***Global Change Impact Studies Centre*** - Details about objectives, manpower, news and events, and publications.
- ***Global Energy Technology Strategy Program*** - Assessing the roles that technology cam play in effectively managing the long-term risks of global climate change since 1998.
- ***Joint Program on the Science and Policy of Global Change*** - Provides links to publications, personnel, forums, sponsor details, job openings and contact information. Based at Massachusetts Institute of Technology, USA.
- ***Kyoto Protocol & Greenhouse Gases*** - This interactive map presents gas emissions in relation to the commitments in the Kyoto Protocol and gives easy access to this information.

- ***Pew Center on Global Climate Change*** - Supports a cooperative approach and brings critical scientific, economic and technological expertise to the global debate on climate change.
- ***Potsdam Institute for Climate Impact Research*** - Interdisciplinary studies in the natural and social sciences, intended to provide a basis for advising policy makers on measures to avoid the causes, mitigate the effects, and adapt to the unavoidable consequences of global change.
- ***Replacing fossil fuels: the scale of the problem*** - Discussion about ways to replace fossil fuels. Information about wind energy, energy storage and energy efficiency is also there.
- ***Research Programme International Climate Policy - HWWA, Hamburg, Germany*** - Scientific/political research on climate policy, emissions trading, Kyoto protocol and greenhouse gas emissions.
- ***Resources for the Future Library: Climate*** - RFF is an NGO based in the U.S. Emphasis on policy studies.
- ***Responding to Climate Change*** - A practical guide to support the United Nations Framework Convention on Climate Change, and the Kyoto Protocol. Provides case studies on greenhouse gas mitigation and adaptation (best accessed using Acrobat).
- ***Shared Spaces*** - Quarterly online magazine by the Netherlands Ministry of Spatial Planning, Housing and Environment. Provides information about various activities of the Ministry, and links for feedback and contact information.
- ***The Tyndall Centre for Climate Change Research*** - UK-based interdisciplinary center seeks to engage with stakeholders and to explore and develop new technologies, life-styles, economic instruments and regulatory mechanisms that will allow climate change to be managed.

- *United Nations Framework Convention on Climate Change (UNFCCC)* - Official site for UN Climate Secretariat: Kyoto protocol, COP4 in Buenos Aires, and world emissions data.
- *US Department of State: Global Climate Change* - Details of official communication on global climate change.
- *Weathervane* - An online forum designed to provide the news media, legislators, opinion leaders, and the interested public with analysis and commentary on U.S. and global policy initiatives related to climate change.
- *World Bank - Global Climate Change* - Information on World Bank activities in relation to climate change, including energy strategy and the Prototype Carbon Fund.
- *WRI: Climate Change Research* - The World Resources Institute identifies opportunities to reduce the risk of global climate change in ways that drive sustainable economic development worldwide. This site provides extensive on-line resources.
- *Climate Protection Strategies for the 21st Century: Kyoto and Beyond* - Report (in English) prepared by the German Advisory Council on Global Change (WBGU) as input to Input to the 9th Conference of the Parties to the Climate Change Convention, 1-12 December 2003, Milano. (December 8, 2003)

8

AGRICULTURE AND BIOTECHNOLOGY

Proponents of free trade argue that liberalization of agriculture will increase food production and improve the economic conditions of farmers around the world. Real life experience, however, is telling a vastly different story as the liberalization of agriculture unfolds. The entry of transnational corporations in the agricultural sector has created devastating conditions in India, leading many small and marginal farmers, for example, to commit suicide due to economic distress.

The entry of transnational corporations in the sector has moved the focus of agriculture from self sufficiency to making profits. On the one hand, reform policies have ended subsidies provided to Indian farmers, leaving them at the mercy of market forces. On the other hand, these policies have increased incentives for large agribusiness companies to farm. Such is the level of disbandment of farmers in India that, as Devinder Sharma, a well known food policy analyst writes:

"In Andhra Pradesh, the state which leads the country in ushering in progressive policies to improve agriculture, chief minister Chandrababu Naidu has launched the Vision 2020 programme that aims to reduce the number of farmers from the present 70 per cent to 40 per cent. While what happens to the remaining 30 per cent of the farming force is none of his concern, the State continues to record the highest suicide rate among the farming community in the country. Such is the government's apathy that instead of resurrecting the policies to help farmers tide over the crisis of livelihood security, chief minister has been talking of deputing psychiatrists to advise farmers not to commit suicide!"

The impacts of liberalization in India are especially important because over 70% of India's population, directly or indirectly, is engaged in the agriculture sector. Add to this the fact the India is home to nearly a third of the world's poor, and one can begin to realize the numbers of people that trade liberalization, especially of the agricultural sector, impacts.

In their pursuit of profits, transnational corporations are eager to push untested technologies and models regardless of the human, environmental and social damages they may cause. India is rapidly moving toward a cash crop economy, at the expense of self-sufficiency. Large multinationals like Monsanto are also pushing the Gene Revolution (biotechnology), as the successor to the Green Revolution, in spite of largely unanswered questions about the impacts of such a technology. There are grand yet unproved claims that biotechnology is the answer to feeding the hungry in the world. Not only are the unanswered questions problematic, biotechnology in agriculture itself is designed so that corporations gain full control over the agricultural system from local farmers.

Ironically, the hunger problem in India is not one of food production but rather food distribution and storage. If left to the free market, as the WTO would like us to do, there will really be no food at all for the poor since we will export all the cash crops and the imported food will be too expensive for the poor to afford!

What is required is food security and achieving food security will involve consciously planning for the needs of the farmers in India, not abandoning them as has been the case since 1991 when the Indian government began to liberalize the economy. It will also involve guaranteeing genuine land reforms for most Indians, especially in the light that land holding size has been steadily declining over the years.

This issue area looks at the problems surrounding liberalization and corporate agriculture in India, and the attempts to turn back the tides of corporate globalization.

9

ENERGY AND CLIMATE CHANGE

Fossil fuels—coal, oil and natural gaspower corporate globalization. India is a case in point.. As it becomes increasingly integrated into the global economy, India is increasing its imports of oil and natural gas, most of it from the Middle East. It is already very reliant on coal for energy. Moving toward an increasing fossil fuel dependence comes with a price both financially as well as in terms of human costs. This Energy and Climate Change Section of our Issue Library explores fossil fuel-based development in India, and alternatives to it.

Investment in fossil fuel-based development comes at the price of a better health care, education, housing and other immediate needs that India has. Whats more, this energy most often does not benefit the vast majority of people in India, but rather just the elite few and the transnational corporations investing in the country. One may not get this picture by visiting Indian cities but one has to remember that over 70% of India lives in rural areas, not urban centers.

For instance, numerous automakers, from Toyota to Mercedes, and even the Indian Tatas, continue to manufacture and market status symbols in the form of gas guzzling Sport Utility Vehicles (SUVs). Yet only a handful of people in India can actually afford one. What about the new airline companies and airports? Again, a small minority of people in India can afford them. We even have instances in places like Singrauli, site of one of the largest thermal power plants, and Dabhol, the site of the infamous Enron power plant, where the poor who live in the proximity of the plant do not have access to power while the government and the corporations are laying grids and pipelines to fuel factories and other foreign investments, while building roadways to distribute their goods. Clearly, energy development in India is for a handful minority that can afford it and who will benefit from it.

This skewed situation is compounded by the injustice of the energy life cycle. Most often, coal mining is done in adivasi (indigenous) areas in India with complete disregard for the human rights of the communities or the ecology. And India has relies heavily on coal; in fact, about 70% of the country's electricity comes from coal. Meanwhile, oil refineries are notorious for air pollution that severely affects human health and the surrounding environment.

And finally, there is the end result of our consumption of fossil fuels. The burning of oil, coal and gas and the resulting release of carbon dioxide is the single largest contributor to climate change. As the climate changes, it is often those that are the poorest and most marginalized that are the hardest hit around the world. India is no different. The cyclones in Orissa in 1999 which resulted in over 10,000 deaths and hundreds of thousand homeless was the kind of intense storm forecast by climate scientists to become more frequent as global warming increasingly becomes a reality; it is a sign of things to come as we ignore the warning signs as a society and continue to depend on fossil fuels for our energy development.

Regardless of what the fossil fuel companies may pay their own scientists to say and write, climate change is happening. We are way past the discussion of whether it might occur, but rather are debating how bad it will be. Needless to say, the global community has no choice but to wean itself away from such a destructive source of energy. It is true that industrialized nations in the North contribute most significantly to climate change with their massive historical and current consumption of fossil fuels. And that they need to lead the transformation away from fossil fuel dependency.

Of course, India has genuine energy needs and we need to meet them. But instead of becoming increasingly addicted to fossil fuels, the country has a tremendous opportunity to buck the trend and develop alternative energies- biomass, solar, wind - renewable energy forms that India has plenty of potential for. Otherwise India too, will be stuck on an increasingly unstable (and finite) supply of oil and gas from the Middle East, and coal at home. If the country does not change its energy strategy, it will begin to look like those industrialized nations that do the dirty business of refining oil on the back of the poor and the marginalized.

And finally, in order to have an equitable distribution of energy, we have to seriously examine who we do business with and how. Deregulation of the power sector, history has shown, increases the cost of power and leads to cases like the Enron fiasco in Dabhol and the price gouging in California. Many of these fossil fuel companies, from Shell to UNOCAL to Chevron, have been complicit in serious abuses of human rights and caused intense ecological degradation. These corporations have no qualms about working with repressive regimes to get their way. And as the Enron disaster is showing us, multinational corporations have far reaching influence in world politics, and can get their way anywhere, including India.

Energy is too precious to be given away to a handful of people for profit

10

GLOBALIZATION

What are the policies that allow for increased corporate investment? How are these policies made and who influences them? Who benefits, who loses? And most importantly, how are movements in India fighting back the impacts and agents of economic reform? This issue area looks at the process of economic globalization, and offers the big picture which is impacting people in India and around the world.

We offer below an excerpt written by Joshua Karliner, on globalization.

What is Globalization? A generation ago, when astronauts first beamed pictures of the Earth back home, the image of a life-sustaining planet floating in infinite space helped inspire social movements to work to save the world from environmental destruction. Today, the stunning portrait of the Earth has also become the icon for a millennial version of manifest destiny known as economic globalization.

We are told that this corporate encirclement of the planet will bring with it greater prosperity, peace and ecological balance. Certainly, it is difficult to resist embracing such an enticing vision, especially when there is no clear alternative on the horizon. In many respects traditional nation-states, including both the high-tech industrial democracies and the multitude of Third World governments, have grown weaker and less relevant. And the community of nations is plagued by growing nationalism and ethnic division leading to xenophobia, fundamentalism, fascist tendencies and war. This leaves the corporate capitalists and the leaders of the industrialized countries to present their neoliberal brand of globalization as inevitable and themselves as healers of the world's ills.

In the absence of a coherent alternative, the transnational corporations carry on inexorably. Increasingly flagless and stateless, they weave global webs of production, commerce, culture and finance virtually unopposed. They expand, invest and grow, concentrating ever more wealth in a limited number of hands. They work in coalition to influence local, national and international institutions and laws. And together with the governments of their home countries in Europe, North America and Japan, as well as international institutions such as the World Trade Organization, the World Bank, the International Monetary Fund and increasingly, the United Nations, they are molding an international system in which they can trade and invest even more freely—a world where they are less and less accountable to the cultures, communities and nation-states in which they operate. Underpinning this effort is not the historical inevitability of an evolving, enlightened civilization, but rather the unavoidable reality of the overriding corporate purpose: the maximization of profits.

The "globalization" we are witnessing today is in fact an acceleration of historical political and economic trends, hastened by the advent of increasingly sophisticated and rapid communications and transportation technologies, the decline of the nation-state (especially in the South), the absence or ineffectiveness of democratic systems of global governance, and the rise of neoliberal economic ideology. Its primary beneficiaries are both the transnational corporations, as well as the privileged consumer classes in the North and to a growing degree, in the industrializing nations of the South.

Excerpted from Joshua Karliner's The Corporate Planet: Ecology and Politics in the Age of Globalization (Sierra Club Books, 1997)

WATER

In the 1970s and 1980s, bottled drinking water in India was unheard of. Now, everyone in the country, it seems, is drinking water from plastic bottles. At least that is how it seems at first glance. Bottled water, controlled by the likes of Nestle, Pepsi and Coca Cola, is mostly out of reach for most Indians, given its prohibitive prices. Most Indians carrying a plastic water bottle are reusing the empty, non-recyclable plastic for carrying tap water rather than consuming bottled water on a regular basis.

The corporate control of water and water distribution in India is increasing. As globalization opens opportunities for private players, investing in water, or and manipulating the scarcity of water, makes increasingly good business sense for corporations.

Indeed, water utility companies like Vivendi and Suez of France have entered the Indian market and are beginning to run water distribution systems. Such systems would provide water to those who can afford the market prices, and those that cannot are out of luck. Clearly, there will be more losers in a country like India than those who stand to gain from such a system.

But it also begs a larger question- who has the right to own water and profit from it anyway? Its not as if these corporations invented water. And while only 4% or so of all the water worldwide is freshwater and fit for drinking, water is part of the global commons, like air; the rights to water should extend to all human beings. The privatization of water for profit will deny a majority of the people in the world a basic right that should be guaranteed.

And to add to the irony of it all, industrial pollution in India is one of the main reasons why so much of our water is not fit to drink. As more and more transnational corporations invest in India, their plants pollute our water systems, increasing the demand for different sources of clean water. As local sources of water become fouled, demand increases for water corporations to sell a product that was once free. Complicit in all this is the World Bank, which promotes these water polluting industries in India in the first place, and then also finances water cleaning projects (that don't work) in the same areas!

Privatization of water is wrong and will not work. The Water Section of the Issue Library looks at the politics of water in India and globally. Water, which has been identified by corporations as the last frontier, clearly needs to be more on the agenda of Indian social movements. So we plan to bring insightful, cutting edge stories on water in order to highlight the problems with the corporate control over water.

11

DEPLETION OF DRINKING WATER: A CASE STUDY OF COCA-COLA

In this study, we the attention of our countrymen and of the whole world to the glaring example of an impending World Danger of Drinking Water in the form of its open commercialization by Coca-Cola, Pepsi and other cold drink companies. In fact the problem is not new to you, neither is peoples sentiment collaboration with the struggle against it, new. Yet we would like you to note some facts concerning the present condition both of the depletion of water and of our struggle against it.

COCA-COLA

The Hindustan Coca-Cola Beverages Private Limited established its bottling plant in Mehdiganj village of Varanasi District of Uttar Pradesh, India, in 1995. Ever since, the people of the place have been constantly protesting against it. The reasons for protest are manifold.

1. Illegal Possession of Village Land

The Company has forcefully occupied more than 33 decimal of village land despite the resistance and notices of the village authorities. District authority and the Court had found them guilty and issued notices to return this public land. But the money power of the company had no troubles in keeping the land against this country's law and people.

The company had also escaped with the theft of stamp duty worth rupees one crore five lakhs and seventy five thousand. Interestingly, the Varanasi court had a judgment passed against them in this regard to pay a penalty of Rs.

3,01,50,000 on 3 March 2003, which the company's underhand power has succeeded to ignore to this day.

2. Polluting Waste

The company had made false publicity among the villagers that their waste would be so pure as edible and manure. But the experiments and experiences have proved to the whole world that not only their waste but the Cola itself is poisonous!, that the polluted water which the company releases contains heavy content of cadmium and lead which has in fact destroyed the entire agricultural crops of the area causing unbearable losses to the village farmers. Surveys have already manifested that various intestinal and skin diseases increased in the surrounding villages during the last five years. Many written complaints on this fact have been made to the administration. Both administration and Pollution Board have made study and discovered the truth of the sicknesses and losses. Despite all these documented facts none of these machineries have ever succeeded to take even a single step to take action against the company, nor to give compensation to the poor villagers.

3. Unjust Management

Coca Cola is one of the highest profit making multinational companies of the world. But the local people have come to experience the cut-throat behaviors of the company to its employees. It has kept all its workers as non permanent. When the employees took steps to unite to demand for just wages, the company dared to use illegal forces, dismissed many and got 13 of its employee leaders implicate in fictitious crimes and court cases for which they are still being troubled.

4. Depletion of Ground Water for Commercial Purposes

The most heinous crime against the people of Mehdiganj and against the whole of our country is the massive exploitation of ground water, more than 250,000 liters of water per day. This colossal use of ground water has dried up open wells and run down the level of bore wells from 30 feet to 50 feet, in this Gangetic Plain so that more than 20 villages are undergoing terrible scarcity of water not only for irrigation but also for drinking! The drought of last two years has worsened this situation leaving people with no alternative than to drive away the exploiters of their drinking water.

5. Poor Response of the Government to these Problems

The government of Uttar Pradesh has been saying that they are concerned about the awful condition of drinking water in the state, and is considering closing deep wells of the government for the sake of the shallow wells of the people.

The fact is that when water is drawn for irrigation only 8 to 10% of the water is used for crops, all the rest 90% returns to the source. But commercialization and chemicalization of water uses the total content of the water drawn out. We have only one question to the UP government, "Is it in a position to stop commercialization of ordinary people's drinking water by Coca Cola"??

A parliamentary committee under the leadership of Hon'ble Sharad Pawar had found that Pepsi and Coke contained heavy contents of poisonous contents after a study on the recommendation of the NGO institution CSE. But all that the whole parliament could do was to remove these soft drinks from their canteens, hinting as if to the whole country that even if the entire nation goes to doom, what concerns to them is their own welfare!!

STRUGGLES AGAINST THE MONSTER OF COCA-COLA

The people of Mehdiganj have been holding peaceful protests against Coca Cola according to the norms of democracy ever since they started to see the deadly dangers it is spreading in the whole area. The Mehdiganj panchayat had unanimously passed a resolution to cancel the license of the company in May 2005. But the state government behaved as a domestic dog of the company and used brutal violence on the peacefully protesting people of the place and on their supporters, beating up men and women alike, locking them up and suing them in the court with fake criminal cases.

Now it has become clear to the people of Mehdiganj and to the whole common man of the country that our government is bought cheap by such companies and their mafia to the extent that people are left with one single option to FIGHT UNTO DEATH, for their own survival. Now people and all People's Movements have determined to fight the decisive battle. Now the government which runs with each hard earned penny of the common people and their vote, will have to make a public decision whether it is with the people or with the companies who fill their pockets. Now is the time to prove that power of justice and truth is still with the people and not with any corrupt politician.

On this occasion we wish to bring the attention of our countrymen and of the whole world to the glaring example of an impending World Danger of Drinking Water in the form of its open commercialization by Coca Cola and Pepsi companies. In fact the problem is not new to you, neither is your collaboration with us in our struggle against it, new. Yet we would like to bring your attention to some facts concerning the present condition both of the depletion of water and of our struggle against it.

So we the people of Mehdiganj have determined to sit on strike in front of the Hindustan Coca-Cola Beverages Private Ltd. bottling plant at Mehdiganj indefinitely until the following demands of our people are fulfilled by the government.

1. Cancel the license of Coca-Cola company
2. Prohibit Coca-Cola company from its daily exploitation of 25 lakh liters of ground water

3. Release the public land illegally occupied by the company from the village panchayat of Mehdiganj
4. Stop the pollution caused by the wastes of Coca-Cola, and compensate the people of Mehdiganj for the losses of life and crops
5. Deposit the dwindled stamp duty of rupees three crores one and half lakh into the treasury immediately
6. Take back all the fake cases into which Coca-Cola has dragged our peaceful protesters, and its employees
7. Ensure a country-wide block on the poisonous soft drinks of Pepsi and Coke

March 23 is the historical day of the victorious martyrdom of Sardar Bhagat Singh, the symbol of daring Indian determination for freedom. This is also the birthday of Dr. Ram Manohar Lohia the symbol of people's revolution in the Free India. March 22, just the preceding day is dedicated for the awareness of Drinking Water of the World. Incidentally related is also the series of World Social Forum gatherings in Pakistan, South Africa and Brazil starting from 23 March. We find no better auspicious day to initiate this indefinite struggle of our people against injustice and exploitation.

Hence may we request you to ensure your participation in this revolution physically with your presence and symbolically with your encouragement, support and contribution.

This presentation was made before the authorities and against coca-cola by the local NGO's and CBO's:

LOK SAMITI, VARANASI
COCA COLA BHAGAO, GAON BACHAO
SANGHARSH SAMITI, MEHDIGANJ
SAJHA SANSKRITI MANCH
NATIONAL ALLIANCE OF PEOPLES MOVEMENTS
AND ALL WELL-WISHERS AND COLLABORATORS

INDEX

A

Abatement of pollution 123, 180
Abolition of child labour 200
Accelerated rural water supply 196
Accountability 18, 30
Acquisition of aircraft 71
Affordability 178
Afforestation 135, 152
Agricultural development strategy 94
Agricultural labour 206
Agriculture productivity 107
Air pollution 82, 126
Airline operation 71
Alkalinity 111
Ambient air quality 83, 84, 85, 127
Animal biotechnology 63
Animal husbandry 95, 108
Anti-sandinista 3
Arable land 111
Arbitrary fashion
Association chajulense 21
Atmospheric concentration 120, 247
Awareness campaign programme 52
Awareness generation 38

B

Balance of payments 43
Benzene gasoline 81
Bharat Jan Gyan Vigyan Jatha 205
Bilateral assistance 37
Biocontrol agents 102
Biodiversity conservation 149
Bio-fertilizers 211
Biological diversity 134, 147
Biological resources 117, 146
Biomass fuels 109
Biomass production 106
Biosensors 63
Biosphere reserves 132
Biotechnology 62, 211, 256
Boreal ecosystem atmosphere study 233
Budgetary constraints 56
Build-own-lease-transfer 75
Building infrastructure 86
Business 202

C

Capacity-building 37, 45, 52, 83, 171
Carbon dioxide 135, 237
Carbon management 235
Carbon mitigation 235
Central Arid Zone Research Institute 154
Central Ground Water Authority 30
Central Pollution Control Board 85, 126
Central Social Welfare Board 192
Chamorro, Violita 2
Chemicalization of water 267
Chlorofluorocarbon 58
Civil Aviation Policy 70, 80
Cleaner Production Technologies 65
Climate 20, 236

Climate exchange 250
Climate monitor 245
Coal Linkage Committee 155
Coca-cola 265
Codal procedures 68
Combating desertification 42, 145, 242
Commercialization 110, 209
Commercialization of water 267
Committee on pricing water 94
Common Minimum Programme 161
Communicable disease 170
Community Health Centres 168
Community Participation 12
Compressed Natural Gas 67, 157
Concept of woman development 186
Conservation 146
Consultation Processes 8
Consumption 47
Contras 2, 3
Cooperation 41, 45, 54, 57, 86, 91, 119, 149, 223
Cooperation international cooperation 144
Cooperation credit structure 118
Cooperative farmers service centre 207
Coordination Committee 14, 90, 106
Coral reefs 133
Corp watch 252
Cost-effective 72
Counter insurrection 8
Credit mechanisms 18
Crop-diversification 18
Cultivate maize 8

D

DALY 171
Decentralization policies 7, 14
Decentralization strategy 199
Decentralized approach 196
Decentralized participatory planning 166
Decentralized information system network 214
Decision-making 27, 48, 197
Deforestation 139
Degradation of forests 109
Democratization 14
Department of energy 237
Depletion of ground water 266
Depreciation allowance 40
Deregulation 66
Desert Development Programme 151
Desertification 150, 154, 242
Destruction of resources 9
Development cooperation 32
Development to empowerment 186
Dignity 194
Directive principles 179
Disability adjusted life years 171
Discretionary controls 31
Dissemination of information 212
District Planning Committee 199
Diversified crop production 22
Diversifying 21
Domestic resources 40
Downstream environments 102
Dredging oporations 104
Drought 150, 242
Drought-prone areas 95
Drylands 243

E

Early Warming Information System 144
Eco-development 147
Economic catchment areas 10
Economic globalization 261
Economic liberalization 110
Economic reforms 34
Ecosystems 92
Effluent treatment plants 225
Employment assurances scheme 196
Empowered committee 120, 121, 131
Energy 155
Energy-efficient technologies 39
Enron Power Plant 258
Entrepreneurial activities 16
Entrepreneurship promotion 17
Enumeration of child labour 200
Environment 250
Environment friendly fuel 50, 67
Environment education 175
Environment impact assessment 44, 128, 129, 182
Environmental protection 29, 39, 88, 182, 183
Environment sustainability 181
Equal Remuneration Act 186
Eradication of Communicable Disease 169
Eradication of poverty 160, 163
Ethanol blended gasoline 82
Evapotranspiration 243
Eventual emancipation 200
Exotic pests 93
Export-Import Policy 32, 43

F

Farm mechanization programme 114, 115
Farmers 22, 114
Farming system 102
Feeder road system 77
Fertilizer production 111
Financial resources 40
Financial sustainability 24
Financing sustainable development 56
Fiscal incentives 49
Floral and faunal resources 148
Fluxes 234
Fly ash 216
Food grain production 107
Forest conservation 92, 125
Forest vegetation 222
Forestry projects 32
Fossil fuel 74, 232, 233, 257

G

Galvanized metal silos 20
Gasoline 81
Gender equity 185
Genetic resources 63
Geosynchronous satellite Launch vehicle 212
Global Dimming 230
Global environment facility 57
Global harmonization 34
Globalization 257
Goal of pollution abetment 49
Green revolution 49
Green house gas 229, 237
Green house warming 231
Ground water extraction 103
Guerrilla movements 7

H

Health 167
Health system 78
Heavy Haul Operations 73

High speed diesel 81
Hot spots 44
Human resource development 114
Human settlement 176
Humanity 108.
Hurricanes 246
Hybrid technology 114

I

INSAT systems 212
Immunodiagnostic tools 63
Independent regulatory system 76
Indian forestry policy 124
Indian remote sensing 219
Indian renewable energy 158
Industrial air pollution 127, 263
Industrialization 203
Industry 65, 202
Information system 38, 53, 113
Infrastructure development 193
Infrastructure facilities 5, 61
Infrastructure for technology 114
Inland water transport 79, 80
Innovative methods 70
Insecticides pest attack 113
Institutional strengthening 35
Institution vacuum 11
Institutionalization 117
Integrated comprehensive scheme 201
Integrated Mission for Sustainable Development 220
Integrated non communicable disease 168
Integrated pest management 101, 107
Integrated transport policy 66, 68
Integration 44
Integration of environmental concept 183
Internal security 3
International Civil Aviation organization 80
International community 185
International cooperation 27, 37
International Geosphere Biosphere Programme 234
International Standardization Organization 202
Investment 16, 40
Irrigation 109
Irrigation schemes 105
Irrigation water 94
Ixil triangle 12

J

Jawahar Rozgar Yojna 178
Joint Forest Management 125

K

Karliner, Joshua 262
Krishi Vigyan Kendras 208

L

LEDA 9, 24
Labour policy 199
Land-based sources 57
Land degradation 111, 243
Land irrigation 108
Legislation 28
Liberalization 33
Liberalization of agriculture 255
Local economic development 9
  Agnecies 10
  Concept 23
  Geographical coverage 12
  Objectives 11
  Organization 13

M

Macro economic stabilization 54
Mahila Samridhi Yojana 187, 18
Management of lakes 103
Marginal farmers 255
Marine biotechnology 62
Marine ecosystems 134
Marine pollution 57
Marine satellite information system 211
Material efficiency 48
Mechanization 112
Medicinal plants 194
Methane Voluntary Programme 238
Ministry of Surface Transport 82
Mitigate climate change 239
Mobilize external resources 15, 25
Moderation of demand 52
Montreal Protocol 35
Morbidity 216
Methyl chloroform 131
Multigauge operations 77
Multilateral environmental agreements 46
Multilateral fund 35, 122, 138
Municipal development organization 14

N

Nagar Palika 196
Naidu, Chandrababu 255
National Agricultural Insurance Scheme 118
National Bureau 64
National Bureau of Plant Genetic Resources 64
National child labour policy 191
National Development Committee 96
National Development Council 198
National Health Programme 167, 169
Notional Information System 53, 113
National Literacy Mission 173
Nation Nutrition Policy 191
National Policy for Children 190
National Policy on tourism 88
National Wasteland Development 216
Nation Water Management Project 105
Natural Resources 22
  Accounting 35
  Aspects 91
  Management 209
Non-agricultural microenterprizes 99
Non-discriminatroy 47
Non-renewable resources 32
Notificaltion 55
Nueva Segovia 22

O

Ocean based resources 57
Open Economy Industrialization 160
Operation back board 172
Over exploitation 32
Overseas economic cooperation fund 78
Ownership 1, 10, 23
Ozone cells 130
Ozone layer 31, 121, 140
Ozone measurement 142

P

PRODERE 5, 6
Panchayati Raj Institution 111, 197
Parliamentary Committees 72

Participatory society 7
Paryavaran Vahinis 123, 138
Pawar, Sharad 267
Peace agreements 2
Pest monitoring promotion 107
Petro-chemical complexes 55
Planning Commission 27, 67, 68, 155
Planning Sustainable development 224
Plant Genetic Resources 116
Plant molecular biology 62
Policia Nacional 3
Polluting waste 266
Polluting control 54, 204
Polymetallic nodules programmes 210, 211
Post-conflict economy 9
Poverty alleviation programme 161
Poverty eradication 36
Primary health care infrastructure 167
Private participation 74
Privatization of water 264
Process of liberalization 33
Production system 19
Programmes 39
Promotive measures 49
Public Transport System 84
Pulse Polio Immunization Campaign 165

R

RFF 251
Railways 76
Ramiro de Leon Carpio 3
Rashtriya Mahila Kosh 187
Recontras 2
Recycling system 51
Reform process 30
Rehabilitation 153
Relocation scheme 226
Removal of poverty 159
Renewable energy programme 136, 158
Renewable hydrogen production 238
Resources mobilization 55
Resources utilization 10
Resources for the future 251
Roads 77
Rural health infrastructure 170
Rural infrastructure fund 118
Rural sports programme 193

S

SDNP 36
STEP 188
SWOT 15
Safe climate 232
Salvadorian government 3
Scheduled tribes finance 195
School curriculum 174
Science 208
Science and technology advisory committee 204, 206
Sexually transmitted disease 170
Shared spaces 253
Sharma, Devinder 255
Shrimp culture industry 30
Small and medium enterprises 60
Social inequity 1
Social infrastructure 177
Social security schemes 201
Social sustainability 23
Soil carbon sequestration 237
Soil salinity 105
Solar photo voltaics 50
Soilid waste 127
Special component plan 195
Special Economics Cooperation Plan 5
Sport utility vehicles 255
State Forest Services 139
State pollution Control Boards 50
Strive for synergy 10
Subsidiary occupation 10

Sulphur 81
Surface 248
Surface transportation 74, 83
Survival economy 51
Sustainable consumption 47
Sustainable development 27, 33, 42, 54, 91
Sustainable Development Network Programme 36, 213, 223
Sustainable tourism 87, 88

T

TIFAC 215
Taj trapezium 226
Target-free approach 164
Task force 62, 68, 69
Technical community 204
Technical sustainability 10, 24
Technology 58
Technology Development Board 65
Technology Vision 59
Temperature trends 248
Thematic mapping organization 227
Therapeutic modality 171
Thermal Power Plant 55, 126
Threat of Global Warming 249
Tinsmith 20
Total exploitable energy 137
Total literacy campaign 173
Tourism 89
Trade 42
Trade Policy Components 30
Trade Unions 199
Traffic efficiency 73
Transferring technology 42
Transformation 259
Transit camps 6
Transnational Corporation 255, 262
Transparency 199
Transport 66
Transportation Technology 262

U

Unidad Revolucionaria Nacional Guatemalteca 4
Universal Elementary Education 172
Universal immunization 164
Unsustainable subsidy 56

V

Village Protection Committee 215
Voluntary Organization 180, 195

W

Waste disposal 50
Waste Minimization Circles 50, 65, 124
Wastelands Development Projects 125
Water 263
Water pollution 218
Water resources consolidation project 105
Water Resources Planning 104
Water scarcity 103
Water logged areas 106
Watershed management 104
Web resources on desertification 244
Wetlands 133
Wild life resources 140
Wind erosion 151

X

Xemamatze 6
Xenobiotics 63